Medical Textile Materials

The Textile Institute and Woodhead Publishing

The Textile Institute is a unique organisation in textiles, clothing and footwear. Incorporated in England by a Royal Charter granted in 1925, the Institute has individual and corporate members in over 90 countries. The aim of the Institute is to facilitate learning, recognise achievement, reward excellence and disseminate information within the global textiles, clothing and footwear industries.

Historically, The Textile Institute has published books of interest to its members and the textile industry. To maintain this policy, the Institute has entered into partnership with Woodhead Publishing Limited to ensure that Institute members and the textile industry continue to have access to high calibre titles on textile science and technology.

Most Woodhead titles on textiles are now published in collaboration with The Textile Institute. Through this arrangement, the Institute provides an Editorial Board which advises Woodhead on appropriate titles for future publication and suggests possible editors and authors for these books. Each book published under this arrangement carries the Institute's logo.

Woodhead books published in collaboration with The Textile Institute are offered to Textile Institute members at a substantial discount. These books, together with those published by The Textile Institute that are still in print, are offered on the Elsevier website at: http://store.elsevier.com/. Textile Institute books still in print are also available directly from the Institute's web site at: www.textileinstitutebooks.com.

A list of Woodhead books on textiles science and technology, most of which have been published in collaboration with the Textile Institute, can be found towards the end of the contents pages.

Related titles

Biotextiles as Medical Implants
(ISBN: 978-1-84569-439-5)

Medical and Healthcare Textiles
(ISBN: 978-1-84569-224-7)

Handbook of Medical Textiles
(ISBN: 978-1-84569-691-7)

Woodhead Publishing Series in Textiles: Number 174

Medical Textile Materials

Yimin Qin

The Textile Institute

ELSEVIER

AMSTERDAM • BOSTON • CAMBRIDGE • HEIDELBERG
LONDON • NEW YORK • OXFORD • PARIS • SAN DIEGO
SAN FRANCISCO • SINGAPORE • SYDNEY • TOKYO
Woodhead Publishing is an imprint of Elsevier

WP
WOODHEAD PUBLISHING

Woodhead Publishing is an imprint of Elsevier
80 High Street, Sawston, Cambridge, CB22 3HJ, UK
225 Wyman Street, Waltham, MA 02451, USA
Langford Lane, Kidlington, OX5 1GB, UK

ISBN: 978-0-08-100618-4 (print)
ISBN: 978-0-08-100624-5 (online)

British Library Cataloguing-in-Publication Data
A catalogue record for this book is available from the British Library

Library of Congress Control Number: 2015948756

For information on all Woodhead Publishing publications
visit our website at http://store.elsevier.com/

Contents

Woodhead Publishing Series in Textiles

1 **Watson's textile design and colour Seventh edition**
Edited by Z. Grosicki
2 **Watson's advanced textile design**
Edited by Z. Grosicki
3 **Weaving Second edition**
P. R. Lord and M. H. Mohamed
4 **Handbook of textile fibres Volume 1: Natural fibres**
J. Gordon Cook
5 **Handbook of textile fibres Volume 2: Man-made fibres**
J. Gordon Cook
6 **Recycling textile and plastic waste**
Edited by A. R. Horrocks
7 **New fibers Second edition**
T. Hongu and G. O. Phillips
8 **Atlas of fibre fracture and damage to textiles Second edition**
J. W. S. Hearle, B. Lomas and W. D. Cooke
9 **Ecotextile '98**
Edited by A. R. Horrocks
10 **Physical testing of textiles**
B. P. Saville
11 **Geometric symmetry in patterns and tilings**
C. E. Horne
12 **Handbook of technical textiles**
Edited by A. R. Horrocks and S. C. Anand
13 **Textiles in automotive engineering**
W. Fung and J. M. Hardcastle
14 **Handbook of textile design**
J. Wilson
15 **High-performance fibres**
Edited by J. W. S. Hearle
16 **Knitting technology Third edition**
D. J. Spencer
17 **Medical textiles**
Edited by S. C. Anand
18 **Regenerated cellulose fibres**
Edited by C. Woodings

19 **Silk, mohair, cashmere and other luxury fibres**
Edited by R. R. Franck
20 **Smart fibres, fabrics and clothing**
Edited by X. M. Tao
21 **Yarn texturing technology**
J. W. S. Hearle, L. Hollick and D. K. Wilson
22 **Encyclopedia of textile finishing**
H-K. Rouette
23 **Coated and laminated textiles**
W. Fung
24 **Fancy yarns**
R. H. Gong and R. M. Wright
25 **Wool: Science and technology**
Edited by W. S. Simpson and G. Crawshaw
26 **Dictionary of textile finishing**
H-K. Rouette
27 **Environmental impact of textiles**
K. Slater
28 **Handbook of yarn production**
P. R. Lord
29 **Textile processing with enzymes**
Edited by A. Cavaco-Paulo and G. Gübitz
30 **The China and Hong Kong denim industry**
Y. Li, L. Yao and K. W. Yeung
31 **The World Trade Organization and international denim trading**
Y. Li, Y. Shen, L. Yao and E. Newton
32 **Chemical finishing of textiles**
W. D. Schindler and P. J. Hauser
33 **Clothing appearance and fit**
J. Fan, W. Yu and L. Hunter
34 **Handbook of fibre rope technology**
H. A. McKenna, J. W. S. Hearle and N. O'Hear
35 **Structure and mechanics of woven fabrics**
J. Hu
36 **Synthetic fibres: Nylon, polyester, acrylic, polyolefin**
Edited by J. E. McIntyre
37 **Woollen and worsted woven fabric design**
E. G. Gilligan
38 **Analytical electrochemistry in textiles**
P. Westbroek, G. Priniotakis and P. Kiekens
39 **Bast and other plant fibres**
R. R. Franck
40 **Chemical testing of textiles**
Edited by Q. Fan
41 **Design and manufacture of textile composites**
Edited by A. C. Long

42 **Effect of mechanical and physical properties on fabric hand**
Edited by H. M. Behery
43 **New millennium fibers**
T. Hongu, M. Takigami and G. O. Phillips
44 **Textiles for protection**
Edited by R. A. Scott
45 **Textiles in sport**
Edited by R. Shishoo
46 **Wearable electronics and photonics**
Edited by X. M. Tao
47 **Biodegradable and sustainable fibres**
Edited by R. S. Blackburn
48 **Medical textiles and biomaterials for healthcare**
Edited by S. C. Anand, M. Miraftab, S. Rajendran and J. F. Kennedy
49 **Total colour management in textiles**
Edited by J. Xin
50 **Recycling in textiles**
Edited by Y. Wang
51 **Clothing biosensory engineering**
Y. Li and A. S. W. Wong
52 **Biomechanical engineering of textiles and clothing**
Edited by Y. Li and D. X-Q. Dai
53 **Digital printing of textiles**
Edited by H. Ujiie
54 **Intelligent textiles and clothing**
Edited by H. R. Mattila
55 **Innovation and technology of women's intimate apparel**
W. Yu, J. Fan, S. C. Harlock and S. P. Ng
56 **Thermal and moisture transport in fibrous materials**
Edited by N. Pan and P. Gibson
57 **Geosynthetics in civil engineering**
Edited by R. W. Sarsby
58 **Handbook of nonwovens**
Edited by S. Russell
59 **Cotton: Science and technology**
Edited by S. Gordon and Y-L. Hsieh
60 **Ecotextiles**
Edited by M. Miraftab and A. R. Horrocks
61 **Composite forming technologies**
Edited by A. C. Long
62 **Plasma technology for textiles**
Edited by R. Shishoo
63 **Smart textiles for medicine and healthcare**
Edited by L. Van Langenhove
64 **Sizing in clothing**
Edited by S. Ashdown

65 **Shape memory polymers and textiles**
J. Hu
66 **Environmental aspects of textile dyeing**
Edited by R. Christie
67 **Nanofibers and nanotechnology in textiles**
Edited by P. Brown and K. Stevens
68 **Physical properties of textile fibres Fourth edition**
W. E. Morton and J. W. S. Hearle
69 **Advances in apparel production**
Edited by C. Fairhurst
70 **Advances in fire retardant materials**
Edited by A. R. Horrocks and D. Price
71 **Polyesters and polyamides**
Edited by B. L. Deopura, R. Alagirusamy, M. Joshi and B. S. Gupta
72 **Advances in wool technology**
Edited by N. A. G. Johnson and I. Russell
73 **Military textiles**
Edited by E. Wilusz
74 **3D fibrous assemblies: Properties, applications and modelling of three-dimensional textile structures**
J. Hu
75 **Medical and healthcare textiles**
Edited by S. C. Anand, J. F. Kennedy, M. Miraftab and S. Rajendran
76 **Fabric testing**
Edited by J. Hu
77 **Biologically inspired textiles**
Edited by A. Abbott and M. Ellison
78 **Friction in textile materials**
Edited by B. S. Gupta
79 **Textile advances in the automotive industry**
Edited by R. Shishoo
80 **Structure and mechanics of textile fibre assemblies**
Edited by P. Schwartz
81 **Engineering textiles: Integrating the design and manufacture of textile products**
Edited by Y. E. El-Mogahzy
82 **Polyolefin fibres: Industrial and medical applications**
Edited by S. C. O. Ugbolue
83 **Smart clothes and wearable technology**
Edited by J. McCann and D. Bryson
84 **Identification of textile fibres**
Edited by M. Houck
85 **Advanced textiles for wound care**
Edited by S. Rajendran
86 **Fatigue failure of textile fibres**
Edited by M. Miraftab

87 **Advances in carpet technology**
Edited by K. Goswami
88 **Handbook of textile fibre structure Volume 1 and Volume 2**
Edited by S. J. Eichhorn, J. W. S. Hearle, M. Jaffe and T. Kikutani
89 **Advances in knitting technology**
Edited by K-F. Au
90 **Smart textile coatings and laminates**
Edited by W. C. Smith
91 **Handbook of tensile properties of textile and technical fibres**
Edited by A. R. Bunsell
92 **Interior textiles: Design and developments**
Edited by T. Rowe
93 **Textiles for cold weather apparel**
Edited by J. T. Williams
94 **Modelling and predicting textile behaviour**
Edited by X. Chen
95 **Textiles, polymers and composites for buildings**
Edited by G. Pohl
96 **Engineering apparel fabrics and garments**
J. Fan and L. Hunter
97 **Surface modification of textiles**
Edited by Q. Wei
98 **Sustainable textiles**
Edited by R. S. Blackburn
99 **Advances in yarn spinning technology**
Edited by C. A. Lawrence
100 **Handbook of medical textiles**
Edited by V. T. Bartels
101 **Technical textile yarns**
Edited by R. Alagirusamy and A. Das
102 **Applications of nonwovens in technical textiles**
Edited by R. A. Chapman
103 **Colour measurement: Principles, advances and industrial applications**
Edited by M. L. Gulrajani
104 **Fibrous and composite materials for civil engineering applications**
Edited by R. Fangueiro
105 **New product development in textiles: Innovation and production**
Edited by L. Horne
106 **Improving comfort in clothing**
Edited by G. Song
107 **Advances in textile biotechnology**
Edited by V. A. Nierstrasz and A. Cavaco-Paulo
108 **Textiles for hygiene and infection control**
Edited by B. McCarthy
109 **Nanofunctional textiles**
Edited by Y. Li

110 **Joining textiles: Principles and applications**
Edited by I. Jones and G. Stylios
111 **Soft computing in textile engineering**
Edited by A. Majumdar
112 **Textile design**
Edited by A. Briggs-Goode and K. Townsend
113 **Biotextiles as medical implants**
Edited by M. W. King, B. S. Gupta and R. Guidoin
114 **Textile thermal bioengineering**
Edited by Y. Li
115 **Woven textile structure**
B. K. Behera and P. K. Hari
116 **Handbook of textile and industrial dyeing. Volume 1: Principles, processes and types of dyes**
Edited by M. Clark
117 **Handbook of textile and industrial dyeing. Volume 2: Applications of dyes**
Edited by M. Clark
118 **Handbook of natural fibres. Volume 1: Types, properties and factors affecting breeding and cultivation**
Edited by R. Kozłowski
119 **Handbook of natural fibres. Volume 2: Processing and applications**
Edited by R. Kozłowski
120 **Functional textiles for improved performance, protection and health**
Edited by N. Pan and G. Sun
121 **Computer technology for textiles and apparel**
Edited by J. Hu
122 **Advances in military textiles and personal equipment**
Edited by E. Sparks
123 **Specialist yarn and fabric structures**
Edited by R. H. Gong
124 **Handbook of sustainable textile production**
M. I. Tobler-Rohr
125 **Woven textiles: Principles, developments and applications**
Edited by K. Gandhi
126 **Textiles and fashion: Materials design and technology**
Edited by R. Sinclair
127 **Industrial cutting of textile materials**
I. Viļumsone-Nemes
128 **Colour design: Theories and applications**
Edited by J. Best
129 **False twist textured yarns**
C. Atkinson

130 **Modelling, simulation and control of the dyeing process**
R. Shamey and X. Zhao
131 **Process control in textile manufacturing**
Edited by A. Majumdar, A. Das, R. Alagirusamy and V. K. Kothari
132 **Understanding and improving the durability of textiles**
Edited by P. A. Annis
133 **Smart textiles for protection**
Edited by R. A. Chapman
134 **Functional nanofibers and applications**
Edited by Q. Wei
135 **The global textile and clothing industry: Technological advances and future challenges**
Edited by R. Shishoo
136 **Simulation in textile technology: Theory and applications**
Edited by D. Veit
137 **Pattern cutting for clothing using CAD: How to use Lectra Modaris pattern cutting software**
M. Stott
138 **Advances in the dyeing and finishing of technical textiles**
M. L. Gulrajani
139 **Multidisciplinary know-how for smart textiles developers**
Edited by T. Kirstein
140 **Handbook of fire resistant textiles**
Edited by F. Selcen Kilinc
141 **Handbook of footwear design and manufacture**
Edited by A. Luximon
142 **Textile-led design for the active ageing population**
Edited by J. McCann and D. Bryson
143 **Optimizing decision making in the apparel supply chain using artificial intelligence (AI): From production to retail**
Edited by W. K. Wong, Z. X. Guo and S. Y. S. Leung
144 **Mechanisms of flat weaving technology**
V. V. Choogin, P. Bandara and E. V. Chepelyuk
145 **Innovative jacquard textile design using digital technologies**
F. Ng and J. Zhou
146 **Advances in shape memory polymers**
J. Hu
147 **Design of clothing manufacturing processes: A systematic approach to planning, scheduling and control**
J. Gersak
148 **Anthropometry, apparel sizing and design**
D. Gupta and N. Zakaria
149 **Silk: Processing, properties and applications**
Edited by K. Murugesh Babu

150 **Advances in filament yarn spinning of textiles and polymers**
Edited by D. Zhang
151 **Designing apparel for consumers: The impact of body shape and size**
Edited by M.-E. Faust and S. Carrier
152 **Fashion supply chain management using radio frequency identification (RFID) technologies**
Edited by W. K. Wong and Z. X. Guo
153 **High performance textiles and their applications**
Edited by C. A. Lawrence
154 **Protective clothing: Managing thermal stress**
Edited by F. Wang and C. Gao
155 **Composite nonwoven materials**
Edited by D. Das and B. Pourdeyhimi
156 **Functional finishes for textiles: Improving comfort, performance and protection**
Edited by R. Paul
157 **Assessing the environmental impact of textiles and the clothing supply chain**
S. S. Muthu
158 **Braiding technology for textiles**
Y. Kyosev
159 **Principles of colour appearance and measurement. Volume 1: Object appearance, colour perception and instrumental measurement**
A. K. R. Choudhury
160 **Principles of colour appearance and measurement. Volume 2: Visual measurement of colour, colour comparison and management**
A. K. R. Choudhury
161 **Ink jet textile printing**
C. Cie
162 **Textiles for sportswear**
Edited by R. Shishoo
163 **Advances in silk science and technology**
Edited by A. Basu
164 **Denim: Manufacture, finishing and applications**
Edited by R. Paul
165 **Fabric structures in architecture**
Edited by J. Ignasi de Llorens
166 **Electronic textiles: Smart fabrics and wearable technology**
Edited by T. Dias
167 **Advances in 3D textiles**
Edited by X. Chen
168 **Garment manufacturing technology**
Edited by R. Nayak and R. Padhye
169 **Handbook of technical textiles Second edition Volume 1**
Edited by A. R. Horrocks and S. C. Anand

170 **Handbook of technical textiles Second edition Volume 2**
Edited by A. R. Horrocks and S. C. Anand
171 **Sustainable apparel**
Edited by R. S. Blackburn
172 **Handbook of life cycle assessment (LCA) of textiles and clothing**
Edited by S. S. Muthu
173 **Advances in smart medical textiles: Treatments and health monitoring**
Edited by L. van Langenhove
174 **Medical textile materials**
Y. Qin

Preface

Medical textile materials are a new discipline encompassing textile technology, material engineering, medical science, and other related areas of science and technology. This is one of the most technically sophisticated and rapidly advancing fields in functional textile materials. Medical textile materials include implant and nonimplant materials, extracorporeal devices, and hygienic textile products. Examples are surgical gauze, operating gowns, protective clothing, hospital bedding, operating towels, absorbent pads, wound dressings, bandages, adhesive plaster, artificial blood vessels, artificial ligaments, surgical sutures, and many other products with unique physical appearances and practical applications. These products are often made by weaving, knitting, braiding, nonwoven, and other novel processing techniques, utilizing a large variety and quantities of basic textile materials such as cotton, viscose rayon, polyester fibers, silk, chitosan fibers, alginate fibers, and other forms of polymeric materials made from polyethylene, polypropylene, polytetrafluoroethylene, polyester, polyamide, polyurethane, and polylactic acid.

In today's textile industry, traditional products are facing increasingly tough competition and there is a need for manufacturers to upgrade their businesses into more value-added areas. As consumers become more and more health conscious, medical textile materials that focus on human health provide a realistic route for the transformation of traditional textile manufacturers. At a time when new knowledge changes people's lifestyles on a regular basis, both technological push and market pull in the medical and healthcare fields have generated a strong momentum for the research, development, manufacture, and sale of novel medical textile materials that provide the inspiration of innovation and scope for business expansion for research organizations and manufacturers alike.

As an emerging multidisciplinary research field, the technical literature on medical textile materials is scattered and there are insufficient textbooks, test standards, product specifications, and other technical materials that are essential for the development of a specialized field. It is therefore important to offer a comprehensive summary of theoretical and practical development in this field which can serve as a platform for people engaged in learning, training, and education about medical textile materials. To provide such a platform, this book set out to introduce the production methods, application areas, industry standards, and test methods for many types of medical textile materials, including superabsorbent materials, functional wound dressings, surgical sutures, artificial ligaments, drug-loaded textile materials, etc. At the same time, this book also introduces the biocompatibility tests, product registration procedures, and research and development strategies that are important for medical-textile-related professionals.

This book can be used for undergraduate teaching in textile engineering, nonwoven engineering, light chemical engineering, biomedical engineering, and polymeric materials. It can also be a source of reference for technical, marketing, and nursing specialists engaged in the research, development, manufacture, and sale of medical devices and pharmaceutical products.

Because medical textile materials cover a large field with a diverse range of specialist knowledge, it is inevitable that this book will not be able to offer precise explanation in all areas, and the author appreciates critical feedback in such cases.

Yimin Qin
2015-4-30

A brief description of the human body

1

1.1 Introduction

The human body is the entire structure of a human organism and consists of almost 100 trillion cells. It is arguably the most complex structure on earth, with billions of microscopic parts, each with its own unique identity, working together in an organized manner for the benefit of the total being. Macroscopically, the human body is a single structure built from billions of smaller structures of four major kinds—cells, tissues, organs, and systems—which are organized biologically to form a whole body comprising head, neck, torso, arms, and legs. The average height of a man is about 1.7–1.8 m, while the height of an average woman is about 1.6–1.7 m. The size, type, and composition of the human body are firstly determined by genes and secondly by postnatal factors such as diet and exercise.

1.2 The systems within the human body

Many diversified activities of men and women are carried out through the combined actions of a complex structure of organ systems distributed in different parts of the human body. These include the musculoskeletal system, cardiovascular system, digestive system, endocrine system, integumentary system, lymphatic system, immune system, respiratory system, nervous system, urinary system, and reproductive system.

Musculoskeletal system: The human skeleton is usually composed of 206 distinct bones, including 26 bones within the spine and vertebral column, eight bones in the cranium, 14 bones in the face, 26 bones in the hyoid, sternum, and ribs, 70 bones in the upper extremities, and 62 bones in the lower extremities. These bones attach to other bones by joints, while the skeletal muscle attaches to the skeleton by tendons. The body is made up of three types of muscle tissue: skeletal, smooth, and cardiac muscles. Each of these has the ability to contract and expand, which allows the body to move and function.

Cardiovascular system: The cardiovascular system comprises the heart, veins, arteries, and capillaries. It is responsible for the circulation of blood, through which oxygen and vital minerals are distributed to the tissues and organs that comprise the body. The main part of the cardiovascular system is the human heart, which is divided into three layers called the endocardium, myocardium, and epicardium, each with a different thickness and function. The left side of the heart is responsible for pumping blood to all parts of the body, while the right side pumps only to the lungs for reoxygenation of the blood.

Medical Textile Materials. http://dx.doi.org/10.1016/B978-0-08-100618-4.00001-7

Digestive system: The digestive system is responsible for processing food and transforming nutrients into energy. It consists of the buccal cavity, esophagus, stomach, small intestine, and large intestine, ending in the rectum and anus. These parts are combined together to form the alimentary canal or the digestive tract.

Endocrine system: The endocrine system is made up of a group of glands that produce the body's long-distance messengers or hormones. Hormones are chemicals that control body functions, such as metabolism, growth, and sexual development. The glands, which include the pituitary gland, thyroid gland, parathyroid glands, adrenal glands, thymus gland, pineal body, pancreas, ovaries, and testes, release hormones directly into the bloodstream, which transports the hormones to organs and tissues throughout the body.

Integumentary system: The integumentary system includes the skin and skin appendages, such as the sweat and sebaceous glands, hair, and nails. As the largest organ system in the human body, it is responsible for protecting the body from most physical and environmental factors.

Lymphatic system: The main function of the lymphatic system is to extract, transport, and metabolize lymph, which is the fluid found in between cells. It is a defense system for the body, filtering out organisms that cause disease, producing white blood cells, and generating disease-fighting antibodies. It also distributes fluids and nutrients in the body, and drains excess fluids and protein so that tissues do not swell. The lymphatic system is made up of a network of vessels that help circulate body fluids. These vessels carry excess fluid away from the spaces between tissues and organs and return it to the bloodstream.

Immune system: This system is responsible for protecting the body from infections and diseases. It is composed of leukocytes, tonsils, adenoids, thymus, and spleen.

Respiratory system: This system comprises the pharynx, larynx, trachea, bronchi, lungs, and diaphragm. It is responsible for inhaling air into the human body and removing carbon dioxide from the body.

Nervous system: The nervous system consists of cells that communicate information about an organism's surroundings and itself. It is made up of the brain, the spinal cord, and nerves. By sending, receiving, and processing nerve impulses throughout the body, the nerve system regulates the muscles and organs to respond to the external environment.

Urinary system: This system includes the kidneys, ureters, bladder, and urethra. It is responsible for eliminating waste from the body, in the form of urine.

Reproductive system: The reproductive system allows humans to produce children. A sperm from the male fertilizes the female's egg, or ovum, in the fallopian tube. The fertilized egg travels from the fallopian tube to the uterus, where the fetus develops over a period of nine months.

1.3 Human anatomy

Human anatomy is concerned with the study of the human body in its macroscopic and microscopic forms. Macroscopically, the morphology of the human body can be seen by the naked eye, and gross anatomy studies anatomical structures concerning organs and organ systems. Microscopic anatomy is the study of minute anatomical structures assisted by microscopes, generating information on tissues and cells in the human body. Histology and cytology are concerned with tissues and cells, respectively.

The history of anatomy has been characterized by continually developing understanding of the functions of organs and structures in the body. Methods in this field have advanced dramatically, from examination of animals through dissection of fresh and preserved cadavers to technologically complex techniques developed in the twentieth century. The studies on human anatomy have generated a comprehensive understanding of the human body, which is important for medical professionals such as physicians, dentists, physiotherapists, nurses, paramedics, and radiographers in their work.

In general, the human body comprises a number of regional groups.

1. Head and neck, including everything above the thoracic inlet, such as brain, eye, pineal gland, pituitary gland, thyroid gland, and parathyroid gland.
2. Upper limbs, including the hand, wrist, forearm, elbow, upper arm, and shoulder.
3. Thorax, which is the region of the chest from the thoracic inlet to the thoracic diaphragm, and contains the heart, lungs, esophagus, thymus gland, and pleura.
4. Human abdomen to the pelvic brim or pelvic inlet, containing the adrenal gland, appendix, bladder, gallbladder, large intestine, small intestine, kidney, liver, pancreatic gland, spleen, and stomach.
5. The back, including the spine and its components, such as the vertebrae, sacrum, coccyx, and intervertebral disks.
6. Pelvis and perineum. The pelvis consists of everything from the pelvic inlet to the pelvic diaphragm. The perineum is the region between the sex organs and the anus.
7. Lower limbs, including everything below the inguinal ligament, including hip, thigh, knee, ankle, and foot.

1.4 Skin

The skin is the outer covering of the human body: it has multiple layers of ectodermal tissue and guards the underlying muscles, bones, ligaments, and internal organs. For the average adult human, the skin has a surface area of 1.5–2.0 m^2, most of it 2–3 mm thick. The average square inch or 6.5 cm^2 of skin holds 650 sweat glands, 20 blood vessels, 60,000 melanocytes, and more than 1000 nerve endings. In addition, the skin provides a rich and diverse habitat for bacteria, which number roughly 1000 species from 19 phyla.

1.4.1 Skin structure

Overall, the skin is composed of three primary layers: the epidermis, which provides waterproofing and serves as a barrier to infection; the dermis, which serves as a location for the appendages of skin; and the subcutaneous layer, also called the hypodermis.

1.4.1.1 Epidermis

The epidermis is the relatively thin, tough, outer layer of the skin, and comprises cornified, translucent, granular, spinous, and basal layers. It contains no blood vessels, and cells in the deepest layers are nourished by diffusion from blood capillaries extending to the upper layers of the dermis. The main types of cells in the epidermis are Merkel cells and keratinocytes, with melanocytes and Langerhans cells also present. Most of the cells in the epidermis are keratinocytes. They originate from cells in the deepest layer of the epidermis, that is, the basal layer, and slowly migrate up toward the surface of the epidermis. Once the keratinocytes reach the skin surface, they are gradually shed and replaced by newer cells pushed up from below. During this process, the cells change shape and composition as they die due to isolation from their blood source. From the basal layer to the corneum, the keratinization process takes place in about 27 days.

1.4.1.2 Dermis

The dermis is the layer of skin beneath the epidermis and is tightly connected to the epidermis by a basement membrane. It harbors many mechanoreceptors and nerve endings that provide the sense of touch and heat. In addition, the dermis contains hair follicles, sweat, sebaceous, apocrine glands, and lymphatic and blood vessels, contained in a thick layer of fibrous and elastic tissue made of collagen, elastin, and fibrillin, which give the skin its flexibility and strength. The sweat glands produce sweat in response to heat and stress, while the sebaceous glands secrete sebum into hair follicles, which keeps the skin moist and soft and acts as a barrier against foreign substances. The blood vessels of the dermis provide nutrients to the skin and help regulate body temperature. Heat makes the blood vessels enlarge (dilate), allowing large amounts of blood to circulate near the skin surface, where the heat can be released. Cold makes the blood vessels narrow (constrict), thereby retaining the body's heat.

1.4.1.3 Subcutaneous layer

The purposes of the subcutaneous layer are to attach the skin to underlying bone and muscle, and to supply it with blood vessels and nerves. It consists of loose connective tissue and elastin. The main cells in the subcutaneous layer are fibroblasts, macrophages, and adipocytes.

1.4.2 Functions of the skin

As it interferes with the external environment, the skin plays a key role in protecting the body against pathogens and excessive water loss. The main functions of the skin include the following.

- Protection: The human skin is an anatomical barrier against pathogens and damage, standing between the internal and external environments to act in bodily defense.
- Water resistance: The human skin acts as a water-resistant barrier so that essential nutrients are not washed out of the body.

- Heat regulation: The blood vessels in the skin can control energy loss through radiation, convection, and conduction. Dilated blood vessels increase perfusion and heat loss, while constricted vessels greatly reduce blood flow and conserve heat.
- Control of evaporation: The human skin provides a relatively dry and semipermeable barrier to fluid loss.
- Esthetics and communication: The condition of the human skin can reflect a person's mood, physical state, and attractiveness.
- Storage and synthesis: The human skin acts as a storage center for lipids and water, as well as a means of synthesis of vitamin D by action of ultraviolet (UV) rays on certain parts of the skin.
- Excretion: Excretion by sweating is a means of self-regulation by the human body. For example, sweat contains urea at a concentration of about 1/130th that of urine.
- Absorption: Cells comprising the outermost 0.25–0.40 mm of the skin absorb oxygen from the environment. In addition, medicine can be administered through the skin using ointments or adhesive patches, such as a nicotine patch or iontophoresis.
- Sensation: The human skin contains a variety of nerve endings that react to heat and cold, touch, pressure, vibration, and tissue injury.

1.4.3 Skin hygiene

Proper skin hygiene is important because unclean skin favors the development of pathogenic organisms. In this respect, the skin supports its own ecosystems of micro-organisms, including yeasts and bacteria, which cannot be removed by any amount of cleaning. Estimates place the number of individual bacteria on the surface of one square inch or 6.5 cm^2 of skin at 50 million, though this figure varies greatly over the average 20 sq ft or 1.9 m^2 of human skin, with oily surfaces such as the face containing over 500 million bacteria per square inch.

On the skin surface, dead cells continually slough off the epidermis, and are mixed with secretions of the sweat and sebaceous glands and dust. If not washed away, this filthy layer on the skin surface undergoes a decomposition process leading to a foul smell. When it is excessively dirty, the normal functions of the skin may be disturbed and it becomes more easily damaged, sometimes leading to skin infections.

1.4.4 Skin care

Proper skin care is vital for its physical appearance. Unhealthy skin tends to appear sallow and dull, whereas a healthy skin appears fully vibrant. In addition, poor skin care can lead to rashes, sores, acne, and wrinkles, with unhealthy skin being more susceptible to disease and infection. In practise, the health of the skin depends on a number of factors. Vitamins play an important role as nutritional elements for the skin. For example, vitamin A, also known as retinoids, benefits the skin by normalizing keratinization, regulating sebum production which contributes to acne, and reversing and treating photodamage, striae, and cellulite. Vitamin D and analogs are used to regulate the cutaneous immune system and epithelial proliferation while promoting differentiation. Vitamin C is an antioxidant that regulates collagen synthesis, forms barrier lipids, regenerates vitamin E, and provides photoprotection. Vitamin E is a membrane antioxidant that protects against oxidative damage and also provides protection against harmful UV rays. As the skin ages, it becomes thinner and more easily damaged. Intensifying this effect is the decreasing ability of the skin to heal itself as a person ages. A validated comprehensive grading scale has categorized the clinical findings of skin aging as laxity (also called sagging), rhytids (also called wrinkles), and the various facets of photoaging, including erythema/telangiectasia, dyspigmentation, solar elastosis, keratoses, and poor texture.

1.4.5 Variability in skin tone

The skin color of different humans is affected by many substances, although the single most important substance is the pigment melanin, which is produced by melanocytes within the skin. In addition to melanin, there are at least five different pigments that determine the overall color of the skin. These pigments are present at different levels in different places. For example, melanin is brown and is present in the germinative zone of the epidermis. Melanoid resembles melanin and is diffused throughout the epidermis. Keratin is yellow to orange in appearance and is present in the stratum corneum and fat cells of dermis and superficial fascia. Hemoglobin is found in blood and develops a purple color, while oxyhemoglobin develops a red color. Individuals with ancestors from different parts of the world can have highly visible differences in skin pigmentation, with sub-Saharan African ancestry tending toward darker skin, while those of northern European descent have fairer skin. Between these extremes are individuals of Asian, Southeast Asian, Native American, Middle Eastern, Polynesian and Melanesian descents.

1.5 Cells

The cell is the smallest unit of life, functioning as the basic structural unit of all known living organisms, including unicellular organisms consisting of a single cell, such as bacteria, and multicellular organisms such as plants and animals. Most plant and animal cells are 1–100 μm in size and are visible under the microscope.

Since the discovery of cells by Robert Hooke in 1665, when he observed cork cells through his microscope, much information has been generated concerning the structures and functions of many varieties of cells. In 1839 Matthias Jakob Schleiden and Theodor Schwann developed the first cell theory, which states that all organisms are composed of one or more cells, that all cells come from preexisting cells, that vital functions of an organism occur within cells, and that all cells contain the hereditary information necessary for regulating cell functions and transmitting information to the next generation of cells. Further studies by generations of scientists around the world have helped humankind understand the sophistications of cells and life.

Cells can be divided into two basic types: prokaryotic and eukaryotic. Prokaryotic cells are usually independent, while eukaryotic cells are often found in multicellular organisms.

1.5.1 Prokaryotic cells

Prokaryotic cells form single-celled organisms, which are the earliest and most primitive forms of life on earth. There are two kinds of prokaryotes, bacteria and archaea, and they are able to live and thrive in various types of environments, including extreme habitats such as hydrothermal vents, hot springs, swamps, wetlands, and the guts of animals. Prokaryotic cells are simpler, and therefore smaller, than eukaryotic cells, lacking a nucleus and most of the other organelles of eukaryotes. They have no true nucleus, as the deoxyribonucleic acid (DNA) is not contained within a membrane or separated from the rest of the cell, but is coiled up in a region of the cytoplasm called the nucleoid. The structure of some bacterial cells includes a capsule that provides an outer covering to protect the cell when it is engulfed by other organisms, assist in retaining moisture, and help the cell adhere to surfaces and nutrients. The cell wall provides the outer covering of most cells, protects the bacterial cell, and gives it shape. Underneath the cell wall, the cell membrane or plasma membrane regulates the flow of substances in and out of the cell and surrounds the cell's cytoplasm, which is a gel-like substance mainly composed of water, and also contains enzymes, salts, cell components, and various organic molecules. On its external surface a bacterial cell usually has pili, which are hair-like structures on the surface of the cell that attach to other bacterial cells, and flagella—long whip-like protrusions that aid in cellular locomotion. The nuclear material of a prokaryotic cell consists of a single chromosome that is in direct contact with cytoplasm. A prokaryotic chromosome is usually a circular molecule, though not forming a nucleus as the DNA is condensed in a nucleoid. Prokaryotes can carry extrachromosomal DNA elements called plasmids, which are usually circular. Plasmids enable additional functions, such as antibiotic resistance.

1.5.2 Eukaryotic cells

Plants, animals, fungi, slime molds, protozoa, and algae are all eukaryotic. These cells are about 15 times wider than a typical prokaryote and can be as much as 1000 times greater in volume. The major difference between prokaryotes and eukaryotes is that eukaryotic cells contain membrane-bound compartments in which specific metabolic activities take place. Most important among these is a cell nucleus, which houses the eukaryotic cell's DNA. This nucleus gives the eukaryote its name, which means "true nucleus."

The eukaryotic DNA is organized in to one or more linear molecules, called chromosomes, which are associated with histone proteins. All chromosomal DNA is stored in the cell nucleus, separated from the cytoplasm by a membrane. Some eukaryotic organelles such as mitochondria also contain some DNA. Many eukaryotic cells are ciliated with primary cilia. Eukaryotes can move using motile cilia or flagella.

1.5.3 Subcellular components

As a cell is an independent living unit, its many components are enveloped inside a membrane which separates its interior from its environment, regulates what moves in and out, and maintains its electric potential. Inside the membrane, most of the cell volume is taken up by a salty cytoplasm containing DNA, the hereditary material of genes, and ribonucleic acid (RNA), with the information necessary to build the various types of proteins such as enzymes. Structurally, the key components of a cell include the following.

Cell membrane: The cell membrane serves to separate and protect a cell from its surrounding environment, and is made mostly from a double layer of lipids and hydrophilic phosphorus molecules; as such, the layer is called a phospholipid bilayer. Embedded within this membrane is a variety of protein molecules that act as channels and pumps to move different molecules in and out of the cell. The membrane is said to be "semipermeable," in that it can either let a substance pass through freely, pass through to a limited extent, or not pass through at all. Cell surface membranes also contain receptor proteins that allow cells to detect external signaling molecules such as hormones.

Cytoskeleton: The cytoskeleton acts to organize and maintain the cell's shape, anchor organelles in place, and help the cellular function during endocytosis. It is also essential for the uptake of external materials by a cell, the separation of daughter cells after cell division, and the movement of parts of the cell during the processes of growth and mobility. The eukaryotic cytoskeleton is composed of microfilaments, intermediate filaments, and microtubules. There are a great number of proteins associated with these, each controlling a cell's structure by directing, bundling, and aligning filaments.

Genetic material: Two different kinds of genetic material exist, DNA and RNA, with the biological information for an organism encoded in its DNA or RNA sequence. RNA is also used for information transport (e.g., mRNA) and enzymatic functions (e.g., ribosomal RNA) in organisms that use DNA for the genetic code itself. Transfer RNA (tRNA) molecules are used to add amino acids during protein translation. A human cell has genetic material contained in the cell nucleus and the mitochondria, with the nuclear genome divided into 23 pairs of linear DNA molecules called chromosomes.

Organelles: Similar to the human body, which contains many different organs such as the heart, lungs, and kidneys, cells also have a set of "little organs" called organelles, which are adapted and/or specialized for carrying out one or more vital functions. Both eukaryotic and prokaryotic cells have organelles, but organelles in eukaryotes are generally more complex and may be membrane bound. There are various types of organelles within a cell.

- **Cell nucleus**: As a cell's information center, the cell nucleus is the place where almost all DNA replication and RNA synthesis (transcription) occur. The nucleus is spherical and separated from the cytoplasm by a double membrane called the nuclear envelope, which isolates and protects a cell's DNA from various molecules that could accidentally damage its structure or interfere with its processing. During processing, DNA is transcribed or copied into a special RNA, called messenger RNA (mRNA), which is then transported out of the nucleus and translated into a specific protein molecule. In prokaryotes, DNA processing takes place in the cytoplasm.
- **Mitochondria and chloroplasts**: Mitochondria are self-replicating organelles that occur in various numbers, shapes, and sizes in the cytoplasm of all eukaryotic cells. Mitochondria play a critical role in generating energy in the eukaryotic cell by oxidative phosphorylation, using oxygen to release energy stored in cellular nutrients such as glucose.

- **Endoplasmic reticulum**: The endoplasmic reticulum (ER) is the transport network for molecules targeted at certain modifications and specific destinations, as compared to molecules that float freely in the cytoplasm. The ER has two forms: rough ER, which has ribosomes on its surface and secretes proteins into the cytoplasm, and smooth ER, which lacks them.
- **Golgi apparatus**: The primary function of the Golgi apparatus is to process and package the macromolecules, such as proteins and lipids, that are synthesized by the cell.
- **Ribosomes**: The ribosome is a large complex of RNA and protein molecules. Each consists of two subunits and acts as an assembly line where RNA from the nucleus is used to synthesize proteins from amino acids. Ribosomes can be found either floating freely or bound to a membrane.
- **Lysosomes and peroxisomes**: Lysosomes contain digestive enzymes (acid hydrolases). They digest excess or worn-out organelles, food particles, and engulfed viruses or bacteria. Peroxisomes have enzymes that rid the cell of toxic peroxides.
- **Centrosomes**: Centrosomes are composed of two centrioles, which separate during cell division and help in the formation of the mitotic spindle. They produce the microtubules of a cell and direct the transport through the ER and Golgi apparatus.
- **Vacuoles**: Vacuoles store food and waste. They are often described as liquid-filled space and are surrounded by a membrane. The vacuoles of eukaryotic cells in plants are usually larger than those in animals.

1.5.4 Functions of the cell

A cell is the basic constituent of a living organism. Its main functions include the following.

Growth and metabolism: Between successive divisions, cells grow through the functioning of cellular metabolism, which is the process by which individual cells process nutrient molecules. Metabolism has two distinct types: catabolism, in which the cell breaks down complex molecules to produce energy and reducing power; and anabolism, in which the cell uses energy and reducing power to construct complex molecules and perform other biological functions. Complex sugars consumed by the organism can be broken down into a less chemically complex sugar molecule called glucose. Once inside a cell, glucose is broken down to make adenosine triphosphate (ATP), a form of energy, through two different pathways. The first pathway, glycolysis, requires no oxygen and is referred to as anaerobic metabolism. Each reaction is designed to produce hydrogen ions that can then be used to make energy packets (ATP). In prokaryotes, glycolysis is the only method used for converting energy. The second pathway, called the Krebs cycle or citric acid cycle, occurs inside the mitochondria and can generate enough ATP to run all the cell functions.

Creation of new cells: Cell division involves a single cell (called a mother cell) dividing into two daughter cells, leading to a growth in multicellular organisms (growth of tissue) and procreation (vegetative reproduction) in unicellular organisms. Prokaryotic cells divide by binary fission, while eukaryotic cells usually undergo a process of nuclear division, called mitosis, followed by a division of the cell, called cytokinesis. DNA replication, or the process of duplicating a cell's genome, is required every time a cell divides. This replication, like all other cellular activities, requires specialized proteins for carrying out the job.

Protein synthesis: Cells are capable of synthesizing new proteins, which are essential for the modulation and maintenance of cellular activities. This process involves the formation of new protein molecules from amino acid building blocks based on information encoded in DNA/RNA. Protein synthesis generally consists of two major steps: transcription and translation. Transcription is the process whereby genetic information in DNA is used to produce a complementary RNA strand, which is then processed to give mRNA and is free to migrate through the cell. These mRNA molecules bind to protein-RNA complexes called ribosomes located in the cytosol, where they are translated into polypeptide sequences. The ribosome mediates the formation of a polypeptide sequence based on the mRNA sequence. The mRNA sequence directly relates to the polypeptide sequence by binding to tRNA adapter molecules in binding pockets within the ribosome. The new polypeptide then folds into a functional three-dimensional protein molecule.

Movement or motility: Cells can move during many processes, such as wound healing, the immune response, and cancer metastasis. For wound healing to occur, white blood cells and cells that ingest bacteria move to the wound site to kill the microorganisms that cause infection. At the same time, fibroblasts (connective tissue cells) move there to remodel damaged structures. In the case of tumor development, cells from a primary tumor move away and spread to other parts of the body. The process is divided into three steps: protrusion of the leading edge of the cell; adhesion of the leading edge and de-adhesion at the cell body and rear; and cytoskeletal contraction to pull the cell forward. Each step is driven by physical forces generated by unique segments of the cytoskeleton.

1.5.5 Different types of cells in the human body

The human body consists of trillions of cells, working together to keep an individual alive. Overall, more than 200 types of cells thrive in the human body, with each cell specially adapted to the particular role it plays. The main types of cells are listed below.

Epithelial cell: Epithelial cells exist throughout the body and function as protective linings covering the insides and outsides of various organs. They can be round, box-shaped, or flattened in structure, and often have cilia on their surfaces to sweep away foreign materials.

Nerve cell: Nerve cells make up the brain, and the spinal cord and all the nerves that branch out of it. The long, thin structure of each nerve cell helps electrical signals pass quickly throughout the body. Many nerve cells wear a layer of fat that helps insulate the signal.

Photoreceptor cell: Rods and cones are two types of specialized nerve cells located in the retina at the back of the eye. They can respond to light and distinguish between different colors.

Bone cell: Although bone tissue contains several different types of bone cells, its main substance is made of a special type of bone cell called an osteocyte, which maintains the regular structure of the bones and regulates the release of calcium into the bloodstream.

Muscle cell: Muscle cells, also known as muscle fibers, are long, spindle-like strands that exist in bundles. Their elongated structure allows the muscle fiber's components, actin and myosin, to slide along each other. This action is necessary for muscle contraction.

Fat cell: Fat cells, also known as adipose cells, store fat droplets in the bulk of their volume for use as energy when the body requires it. They can be found surrounding internal organs and just under the skin.

Red blood cell: Tiny, round, disk-shaped red blood cells constantly transport oxygen throughout the body. Their unique structure allows them to squeeze through even the tiniest of spaces, efficiently carrying oxygen to every other cell in the body.

Macrophage: Macrophages are also known as white blood cells. They roam freely in the blood scouting for harmful bacteria or other foreign particles, which they readily destroy by engulfing them and breaking them down.

Neutrophil: Neutrophils are a type of phagocyte and are normally found in the bloodstream. During the early phase of inflammation, particularly as a result of bacterial infection, environmental exposure, and some cancers, neutrophils are one of the first-responders of inflammatory cells to migrate toward the site of inflammation.

Mast cell: A mast cell is derived from the myeloid stem cell and is part of the immune system; it contains many granules rich in histamine and heparin. Although best known for their role in allergy and anaphylaxis, mast cells play an important protective role as well, being intimately involved in wound healing and defense against pathogens.

Fibroblast: A fibroblast is a type of cell that synthesizes the extracellular matrix and collagen, the structural framework for animal tissues, and plays a critical role in wound healing. Fibroblasts are the most common cells of connective tissue in animals.

1.6 Summary

This chapter offers a brief description of the main features of the human body and shows the areas of the body where medical textile materials may be used. The chapter describes the various systems within the human body, and analyzes the anatomical structure of the body as a whole and the skin and cell structures in detail. The physical composition of the human body is illustrated at four levels: cells, tissues, organs, and systems.

Sources for further information

[1] Lyons I. Biomedical science: lecture notes. Oxford: Wiley Blackwell; 2011.

[2] Sussman C, Bates-Jensen B. Wound care: a practice manual for health professionals. Baltimore (MD): Wolters Kluwer; 2012.

[3] Proksch E, Brandner JM, Jensen JM. The skin: an indispensable barrier. Exp Dermatol 2008;17(12):1063–72.

[4] Madison KC. Barrier function of the skin. J Invest Dermatol 2003;121(2):231–41.

[5] Grice EA, Kong HH, Conlan S, et al. Topographical and temporal diversity of the human skin microbiome. Science 2009;324(5931):1190–2.

[6] Anthea M, Hopkins J, McLaughlin CW, et al. Human biology and health. New Jersey: Prentice Hall; 1993.
[7] Shapiro SS, Saliou C. Role of vitamins in skin care. Nutrition 2001;17(10):839–44.
[8] Orgel LE. The origin of life: a review of facts and speculations. Trends Biochem Sci 1998;23(12):491–5.
[9] Griffiths G. Cell evolution and the problem of membrane topology. Nat Rev Mol Cell Biol 2007;8(12):1018–24.
[10] Vickaryous MK, Hall BK. Human cell type diversity, evolution, development, and classification with special reference to cells derived from the neural crest. Biol Rev 2006;81:425–55.

An overview of medical textile products

2

2.1 Introduction

Medical textiles are fiber-based products and structures used in a medical environment either for the treatment of an injury or for the provision of an appropriate condition in the clinical treatment of a wound or an illness. They embrace all those textile materials used in healthcare and hygiene applications. Medical textiles belong to a group of products generally referred to as "technical textiles," which is a term coined in the 1980s to describe the growing variety of products and manufacturing techniques being developed primarily for their technical properties rather than their appearance or other esthetic characteristics. To reflect the growing importance of technical textiles, a major international exhibition, Techtextil, was launched in 1985 at Messe Frankfurt Exhibition GmbH, where a simple taxonomy was developed and has been used ever since to describe the scope of this new industry and the market sector. It comprises a wide range of products, including the following.

Agrotech (agro textiles): Textiles used in agriculture are termed agro textiles. They are used for crop protection, fertilization, and other agriculture-related purposes. The essential properties required are strength, elongation, stiffness, biodegradation, resistance to sunlight, and resistance to toxic environments. All these properties help with the growth and harvesting of crops and other foodstuffs. There is a growing interest in using materials which gradually degrade.

Mobiltech (automotive and aerospace textiles): These textiles are used in the manufacture of automobiles and aircraft. For example, carbon-fiber-reinforced composites are used in the manufacture of airplane parts, while carbon fiber is used for making higher-end tires. Nylon has a high bursting strength and is used for making air bags in cars. Other examples include spacesuits, parachutes, inflatable boats, seat covers, safety belts, etc.

Buildtech (construction textiles): These are textile materials used in construction, such as concrete reinforcement, facade foundation systems, insulation, proofing materials, air conditioning, noise prevention, visual protection, building safety, etc.

Clothtech (clothing textiles): Clothtech refers to technical textiles for clothing applications, especially in a finishing process where fabric is treated under pressure and high temperatures, in which case the technical textile supports the fabric for smooth processing.

Geotech (geo-textiles): Geo-textiles are usually permeable fabrics used in combination with soils in civil engineering, earth and road construction, dam building, soil sealing, and drainage systems. These fabrics should have high strength and low moisture absorption, and should be durable. Examples include nonwoven and woven fabrics made from glass, polypropylene, and acrylic fibers.

Medical Textile Materials. http://dx.doi.org/10.1016/B978-0-08-100618-4.00002-9

Hometech (domestic textiles): These are textiles used in a domestic environment, such as interior decoration and furniture, carpeting, curtains, cushion materials, fireproofing, floor, and wall coverings.

Indutech (industrial textiles): These are textiles used in chemical, electrical, and mechanical engineering, such as screen printing, filtration, lifting and conveying equipment, sound proofing, roller covers, cable insulation, fuel cells, etc.

Ekotech (environmentally friendly textiles): These include new applications for textiles in environmental protection, such as erosion protection, air cleaning, prevention of water pollution, waste treatment, and recycling.

Packtech (packaging textiles): These include packaging materials, silos, containers, bags, canvas covers, marquee tents, etc.

Sporttech (sports textiles): These include shoes and sports equipment for flying, sailing, climbing, angling, and cycling.

Protech (protective textiles): These products can offer protection against extreme cold, radiation, harmful particulate matter, harmful chemicals, extreme heat and/or fire, molten metals, bacterial and viral contamination, ballistic and mechanical threats, etc., where high-performance fibers are often used. The current size of the personal protective clothing and equipment market in the European Union is estimated to be about €10 billion. Advances in novel specialty fibers, nanoparticles, and the integration of microelectronic components into fabrics and garments have opened up new frontiers in the development of textile-based materials for personal protection.

Cosmetotextiles (cosmetic textiles): Cosmetic textiles are an industry that has grown along with consumer interest in wellness and well-being. It involves the use of textile materials to deliver a wide range of microencapsulated ingredients such as aloe vera, vitamin E, retinol, and caffeine that can offer moisturizing, firming, or slimming benefits. The next generation of cosmetic textile products could potentially go beyond beauty, by utilizing innovative new methods to deliver medical, antiaging, and stress-relieving benefits through apparel textiles and other products. The name "cosmetotextiles" has been coined to designate textiles with cosmetic properties, although these types of textiles can deliver a wide range of bioactive agents, such as mosquito repellents, odor reducers, antimicrobials, and ultraviolet-protection agents.

Within each of these headings are literally hundreds of products and applications for textiles, some traditional, some replacing other well-established materials and techniques, and some that have been newly created by the unique properties and capabilities of textile materials and structures.

2.2 Characteristics of medical textiles

Medical textiles are an important part of the large variety of technical textile products, ranging from high-volume disposable products for baby diapers, feminine hygiene, and adult incontinence through to extremely specialized and high-value textile products for use in blood filtration, surgical sutures, prostheses, and, most recently, scaffolds for new tissue growth.

From a scientific point of view, medical textiles are located at the interfaces between technical disciplines and life sciences. On the one hand, the technical aspect concerns textile engineering, material chemistry, process control, testing and certification, etc., which are needed for the manufacture of high-quality medical textile products. On the other hand, life sciences such as medicine, microbiology, and other related subjects are required for the development of functional performances of these products. In practice, the different scientific and technical disciplines interact and overlap with one another, with new developments in any one of these branches able to generate new innovations in others. For example, new superabsorbent and gel-forming substances invented in chemistry have been applied in the development of new baby diapers and adult incontinence products.

Medical textiles are also closely linked with other applications for technical textiles. For example, protective textiles used in the operating theater are similar to other types of protective clothing, such as those used for chemical, biohazard, or foul-weather protection, and new developments intended for foul-weather protective clothing, such as densely woven microfiber fabrics or breathable laminated textiles, have been used in surgical gowns and drapes. As a consequence of the progress made in the many scientific and technical disciplines related to medical textiles, surgeons, nurses, operating-room staff, and, most importantly, patients can achieve a high level of protection as well as a good wearing comfort during medical treatment.

2.3 Constituent elements of medical textile products

Textile materials used for medical applications include fibers, yarns, fabrics, and many types of composites, which are processed in many ways to form the medical textile products. As shown in Figure 2.1, the basic constituent for medical textile products is polymers, which are made into fibers through a fiber-making process. These polymers differ from those used in conventional textiles, in that they should possess nontoxicity, nonallergenic response, the ability to be sterilized, good mechanical properties, strength, elasticity, durability, and biocompatibility. Sterility is an essential requirement for medical textile products, and therefore the polymers must be able to withstand the harsh physical and chemical conditions that are generally found in a sterilization process.

Depending upon the applications, the major requirements for medical textiles are absorbency, tenacity, flexibility, softness, biocompatibility, and at times biostability or biodegradability. These requirements are met by using appropriate polymers processed into fibers and textiles that possess the right structures to meet the needs of various end users.

Fibers used in the medical textile industry may vary from natural fibers such as cotton, silk, and regenerated wood fluff to synthetic fibers such as polyester, polyamide, polyethylene, polypropylene, and glass fiber. Table 2.1 shows the applications of different types of fibers in medical textile products.

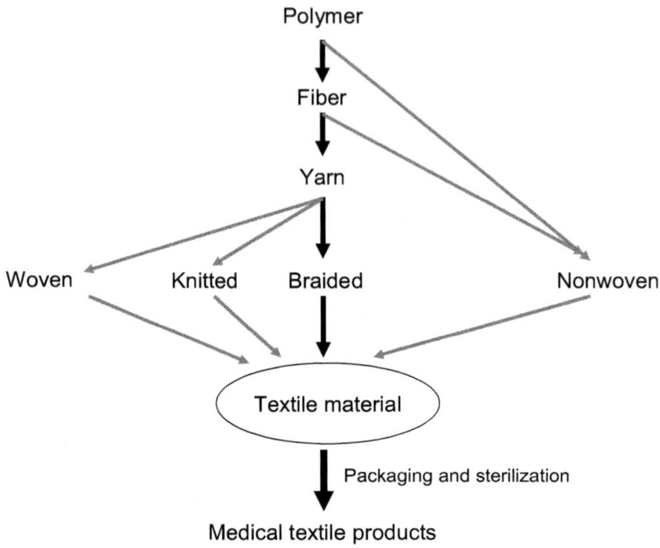

Figure 2.1 The structural hierarchy of medical textile products.

Table 2.1 Applications of different types of fibers in medical textile products

Fiber type	Applications
Cotton	Surgical gowns, bedding, sheets, pillow covers, uniforms, surgical hosiery
Viscose	Caps, masks, wipes
Polyester	Gowns, masks, surgical cover drapes, blankets, cover stock
Polyamide	Surgical hosiery
Polypropylene	Protective clothing
Polyethylene	Surgical covers, drapes
Glass	Caps, masks
Elastomer	Surgical hosiery

For the production of medical textile products for healthcare and hygiene, a number of crucial issues have been identified and debated among clinicians, environmentalists, and pharmaceutical companies for a long time. These issues include:

- natural versus chemical fibers for the manufacture of textile materials;
- disposable versus reusable or durable fabrics for the production of textile products;
- antimicrobial fibers versus antimicrobial finishing or coating for infection control;
- methods of disposal of clinical waste such as landfills versus incineration or other forms of medical waste disposal.

There is a general move toward an increased use of natural polymers that are biocompatible, biodegradable, and nontoxic.

2.4 Classification of medical textile products

Medical textile products can be divided into four groups depending upon the usage: healthcare and hygiene products, extracorporeal devices, implantable materials, and nonimplantable materials.

2.4.1 *Healthcare and hygiene products*

The healthcare and hygiene sector is important for medical textile products. Vast numbers of healthcare and hygiene products are typically used either in the operating theater or in hospital wards for the hygiene, care, and safety of staff and patients. They can be washable or disposable.

Healthcare and hygiene products used in an operating theater include surgical gowns, caps and masks, patient drapes, and cover cloths of various sizes. During surgical operations, it is essential to maintain a clean environment in the operating theater and a strict control of infection. A possible source of infection to the patient is the pollutant particles shed by the nursing staff, so surgical gowns should act as a barrier to prevent the release of pollutant particles into the air. Traditional surgical gowns are woven cotton goods that not only allow the release of particles but also are a source of contamination themselves, because of the generation of a high level of dust. Disposable nonwoven surgical gowns are often made of composite materials composed of nonwoven fabrics and polyethylene films, which can prevent contamination to patients.

In addition to gowns, surgical masks and caps are also used to prevent contamination. In order to be comfortable for the surgeons, they should have a high-filter capacity and a high level of air permeability, as a well as being lightweight and nonallergic. Surgical drapes and cover cloths are used to cover patients and working areas around the patients. They should be completely impermeable to bacteria and absorbent to body perspiration and wound exudate.

Medical textile products used in hospital wards include bedding, clothing, mattress covers, incontinence products, clothes, and wipes. Incontinence products for patients are available in both diaper and flat-sheet forms, with the latter being used for bedding. Cloths and wipes are made from tissue paper or nonwoven-bonded fabrics, which may be soaked with an antiseptic finish.

Superabsorbent fibers are widely used in healthcare and hygiene products. These fibers can absorb up to 50 times their mass in water, whereas conventional wood-pulp and cotton linter absorbents absorb approximately six times their mass in water. These superabsorbent fibers are also advantageous compared to superabsorbent powders in that, due to their small diameter ($\approx 30\,\mu m$) and high surface areas, they typically absorb 95% of the ultimate capacity within 15 s.

Table 2.2 shows some examples of healthcare and hygiene products.

Table 2.2 **Examples of healthcare and hygiene products**

Products	Fiber type	Fabric type
Surgical gowns	Cotton, polyester, viscose rayon, polypropylene	Nonwoven, woven
Caps, masks	Viscose rayon, glass, polyester	Nonwoven
Surgical covers, drape cloths	Polyester, polyethylene	Woven, nonwoven
Bedding, blankets, sheets, pillow covers	Cotton, polyester	Woven, knitted
Uniforms, protective clothing	Cotton, polyester, polypropylene	Woven, nonwoven
Incontinence diaper sheets with cover stock, absorbent layer, and outer layer	Polyester, polyethylene, polypropylene, wood fluff, superabsorbents	Nonwoven
Wipes	Viscose rayon	Nonwoven
Surgical hosiery	Polyamide, polyester, cotton, elastomeric yarns	Nonwoven, knitted

2.4.2 *Extracorporeal devices*

Extracorporeally mounted devices are used to support the function of vital organs, including artificial kidneys (dialyzers), artificial livers, and mechanical lungs. An artificial kidney is a tiny instrument about the size of a two-cell flashlight, made with hair-sized hollow cellulose fibers or hollow polyester fibers, which is used to remove waste products from patients' blood. An artificial liver is made of hollow cellulose fibers to separate and dispose of patient plasmas and supply fresh plasma. A mechanical lung is made with hollow polypropylene fibers or a hollow silicone membrane, which is used to remove carbon dioxide from patients' blood and supply fresh oxygen. An artificial heart is a plastic pump lined with decomposed velour to reduce damage to blood, and is a chambered apparatus about the size of a human heart. Silastic backing makes the fabric impervious to emerging gas, which is not desirable in the blood.

2.4.3 *Implantable materials*

Textile fibers, yarns, fabrics, composites, and three-dimensional shape fabrics made from woven, knitted, nonwoven, braided, and embroidered materials play a vital role in the manufacture of various implants, including the replacement of diseased or nonfunctioning blood vessels and segments of the aorta or other big arteries. Technically, it is even feasible to produce vascular prostheses as fine as 2–3 mm in diameter. These materials are used in effecting repair to the body, whether it is wound closure using sutures or in replacement surgery such as vascular grafts and artificial ligaments.

Biocompatibility is of prime importance if textile materials are to be accepted by the body. In this respect, four key factors will determine how the body reacts to the implant.

Table 2.3 **Examples of implantable materials**

Products	Fiber type	Fabric type
Sutures	Collagen, polylactide, polyglycolide, polyester, polyamide, polytetrafluoroethylene, polypropylene, polyethylene	Mono-filament, braided
Artificial tendons, artificial ligaments, artificial skin, artificial lumen, eye contact lenses	Polytetrafluoroethylene, polyester, polyamide, silk, carbon, etc.	Woven, braided
Orthopedic implants, artificial joints, artificial bones	Silicone, polyacetal, polyethylene, polysulphone, carbon, polyester, glass, ceramic	Composite
Cardiovascular implants, vascular grafts, heart valves	Polyester, polytetrafluoroethylene	Knitted, woven

- The most important factor is porosity, which determines the rate at which human tissue will grow and encapsulate the implant.
- Small circular fibers are better encapsulated within human tissue than larger fibers with irregular cross-sections.
- The fiber and its constituent polymer must not release toxic substances, and the fiber should be free from surface contaminants such as lubricants and sizing agents.
- The fiber should be biodegradable.

Textile materials are particularly suited for implants because they are usually biocompatible materials that can be made into two- or three-dimensional structures, with mechanical characteristics adapted to the environment. The properties can also be adjusted by controlling the macroscopic structure of the textile materials and by specific surface design.

Table 2.3 shows some examples of implantable materials.

A vascular prosthesis is a typical implant, and was first produced from the polyamide fiber in 1956. The prosthesis can be made from a variety of synthetic materials, including polyester, PTFE, polypropylene, and polyacrylonitrile fibers. Polyester and PTFE fibers are commonly used for the manufacture of vascular prostheses.

The major requirements of a good vascular graft include nonfraying, flexibility, durability, biocompatibility, stability to sterilization, and resistance to bacteria/viruses. A knitted polyester vascular prosthesis has become the standard vascular graft for replacement of arterial vessels of 6 mm and greater.

2.4.4 Nonimplantable materials

Nonimplantable materials are used for external applications on the body and may or may not make contact with the skin. They include wound dressings, bandages, plasters, pressure garments, orthopedic belts, etc. Table 2.4 shows some examples of nonimplantable medical textile materials.

Table 2.4 **Examples of nonimplantable medical textile materials**

Products	Fiber type	Fabric type
Absorbent pads, wound contact layer	Cotton, viscose, silk, polyamide, polyethylene	Nonwoven, knitted, woven
Simple bandages, elastic bandages	Cotton, viscose, polyamide, elastomeric fiber and yarn	Woven, knitted, nonwoven
Plasters	Viscose, polyester, polypropylene, perforated film	Knitted, woven, nonwoven
Gauze	Cotton, viscose	Woven, nonwoven
Wadding	Viscose, cotton, wood-pulp	Nonwoven

Surgical dressings are one of the main types of nonimplantable medical textile products, and are used to cover, protect, and support the injured or diseased body part, and to absorb fluid from wounds. There are five different types of materials.

- Primary wound dressing.
- Absorbent pad.
- Bandage.
- Protective layer.
- Adhesive tape.

Primary wound dressings are placed next to the wound surface. They are required to possess the following properties.

- Healing properties, regulated mainly with substances which are applied or added to the dressing.
- Causing no mechanical injury to a granulating wound.
- Low adherence.
- Stable spatial structure.
- Easy penetration of wound exudate into the absorbing dressing.
- Painless changing of the dressing.

Traditional surgical gauzes are made of loosely woven cotton yarns, while surgical bandages are made of narrow fabrics either woven or knitted from cotton or linen. The various types of bandages often have different functions in different applications. For example, simple bandages are used to hold dressings in place. Elasticated bandages impart support and comfort, while light support bandages help manage sprains or strains. Compression bandages assist in the treatment and prevention of deep vein thrombosis and leg ulceration. Orthopedic bandages are used under plaster casts to provide padding and prevent discomfort.

2.5 Summary

In recent years, technical textile materials have found many different kinds of applications in the healthcare sector. The diversity of applications encountered in medical and healthcare products is remarkable, ranging from simple bandages, biocompatible implants and tissues, and antimicrobial wound treatment materials to prosthetics and intelligent textiles. This sector offers the greatest scope for the development of the most sophisticated and highest-value textiles for niche applications. Many end users with their disparate requirements create opportunities for all kinds of textile materials, such as fibers, mono- and multifilament yarns, and woven, knitted, nonwoven, braided, and composite fabrics.

Starting from plain cotton-based products, medical textiles have seen rapid development over the last few decades. Nowadays, new biodegradable fibers have enabled the development of novel types of implants, and modern textile machines can produce three-dimensional spacer fabrics that give superior performance over traditional textile materials. These and many other advances have made medical textiles an essential element in modern disease management, and they are becoming more and more important with the increasing number of elderly people in the populations of developed countries.

As the medical textile industry grows both in size and importance, it has become increasingly necessary to develop a comprehensive and systematic understanding of the science and technology involved in the research, development, manufacture, and applications of medical textile materials, where knowledge, expertise, and interdisciplinary collaboration between medical researchers, microbiologists, physiologists, and textile scientists are required. The present book aims to offer readers in many different areas an overall introduction to the basic sciences, technologies, product varieties, technical characteristics, government regulations, etc. that are required in the research, development, and application of medical textile materials.

Sources for further information

[1] Bartels VT, editor. Handbook of Medical Textiles. Cambridge: Woodhead Publishing Ltd; 2011.

[2] Czajka R. Development of medical textile market. Fibres Text East Eur 2005;13(1):13–5.

[3] Qin Y. Advanced wound dressings. J Text Inst 2001;92(2):127–38.

[4] Anand S, editor. Medical Textiles. Cambridge: Woodhead Publishing Ltd; 2001.

[5] Desai A. Special textiles used for manufacturing healthcare and hygiene products. Text Mag May 2003:73–6.

[6] Rajendran S, Anand S. Contribution of textiles to medical and healthcare products and developing innovative medical devices. Indian J Fibre Text Res 2006;31:215–29.

[7] Brasch J, Rucker M, Haise C. Medical textiles that suit the user: predicting health care workers' preference for disposable versus reusable surgical gowns. Health Mark Q 2013; 30(2):162–76.

[8] Penn LS, Gillis PP, Lay RN. A geometric approach to fibrous structures for biological soft tissue replacement. Text Res J 2001;71(2):93–9.

[9] Frank KK, Laurencin CT. 3-D textile scaffolds for tissue engineering. Text Asia May 2000:27–30.

[10] Malik A. Polymers and fibres in disposable medical products. Asian Text J 2001; 10(1):36–40.

[11] Katzer K. Polyethylene polymers for hygiene market. Asian Text J 2002;11(6):30–4.
[12] Qin Y. Novel antimicrobial fibres. Text Mag 2004 (2):14–17.
[13] Viju S. Biodegradable polyesters for medical applications. Indian Text J June 2008:75–9.
[14] Overcash M. A comparison of reusable and disposable perioperative textiles: sustainability state-of-the-art 2012. Anesth Analg 2012;114(5):1055–66.
[15] Rutala WA, Weber DJ. A review of single-use and reusable gowns and drapes in health care. Infect Control Hosp Epidemiol 2001;22(4):248–57.
[16] Qin Y. Functional alginate fibers. Chem Fibers Int 2010;3:32–3.
[17] Patel SR, Urech D, Werner HP. Surgical gowns and drapes into the 21st century. Br J Theatre Nurs 1998;8(8):30–2.
[18] Petrulyte S. Advanced textile materials and biopolymers in wound management. Dan Med Bull 2008;55(1):72–7.

A brief description of textile fibers

3

3.1 Introduction

A textile fiber is a unit of matter, either natural or manufactured, that forms the basic element of fabrics and other textile structures. A fiber is characterized by having a length at least 100 times its diameter or width. Fibers can be spun into yarns or made into fabrics by various methods, including weaving, knitting, braiding, felting, and twisting. The essential requirements for fibers to be spun into yarns include a length of at least 5 mm, flexibility, cohesiveness, and sufficient strength. Other important properties are elasticity, fineness, uniformity, durability, and luster.

3.2 Fiber-forming polymers

Polymers are the basic constituent of textile fibers. A polymer is a large molecule composed of repeating structural units which are connected by covalent chemical bonds to form a long molecular chain. For example, cellulose is a natural polymer with each single molecule comprising hundreds of glucose monomers, while polyethylene is made of repeating units based on ethylene monomers. Polymers encompass a large class of compounds comprising both natural and synthetic materials with a wide variety of properties. Because of the extraordinary range of properties of polymeric materials, they play an essential and ubiquitous role in everyday life, ranging from synthetic plastics and elastomers to natural biopolymers such as nucleic acids and proteins that are essential for life. The three main forms of polymeric materials are plastics, rubber, and fibers, which can be made into a vast number of end products.

3.2.1 Polymer structures

A polymer is characterized by its molecular and supramolecular structures. On the molecular level, the properties of polymeric materials are influenced mostly by the chemical composition of the polymer and its constituent monomers. In addition, in its condensed state the conformation and configuration of the polymeric chains are important factors in the polymer structure. On the supramolecular level, the properties of polymeric materials are further influenced by the three-dimensional arrangement of the polymeric chains. Some of the key issues concerning polymer structures are discussed below.

Monomers and repeating units: Monomers are the building blocks of a polymeric chain, and polymers are often named according to the type of the constituent monomer residues or repeating units. For example, the generic name for cellulose is poly-(1,4-β-D-glucose), based on the fact that it is derived from D-glucose units linked

Medical Textile Materials. http://dx.doi.org/10.1016/B978-0-08-100618-4.00003-0

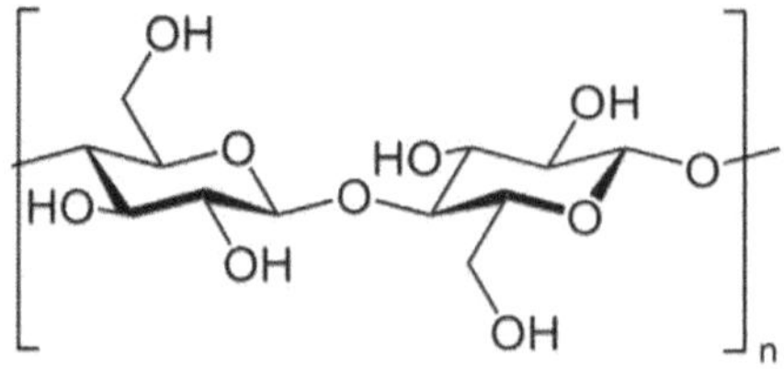

Figure 3.1 The chemical structure of cellulose.

through β(1→4)-glycosidic bonds, as shown in Figure 3.1. Polymers with a single type of repeating unit are known as homopolymers, while those containing a mixture of repeating units are termed copolymers. Biological polymers such as proteins and polysaccharides are often composed of a variety of different but structurally related monomer residues. For example, polynucleotides such as DNA are composed of a variety of nucleotide subunits. A polymer molecule containing ionizable subunits is known as a polyelectrolyte.

Chain length and molecular weight: The number of monomers in a polymer chain is known as the degree of polymerization, and this is one of the most important structural parameters of a polymeric material. Long chains with associated high molecular weights are related to high strength and toughness, low chain mobility, and a high glass transition temperature, as a result of the increase in chain interactions through van der Waals attractions and chain entanglements. These interactions tend to fix the individual chains more strongly in position and resist deformation and matrix breakup at both higher stresses and higher temperatures. Technically, the size of a polymer chain is quantified by the degree of polymerization and molecular weight. Since synthetic polymerization techniques typically yield a polymer product with a range of molecular weights, the weight is often expressed statistically to describe the distribution of different chain lengths in the polymer. The "number average molecular weight" and "weight average molecular weight" are commonly used to characterize the molecular weight of a polymer, with the ratio of these two values, that is, the polydispersity index, used to express the "width" of molecular weight distribution.

Monomer arrangement in copolymers: In a copolymer, the different types of monomers are distributed along the backbone in a number of ways. In a random copolymer the monomers distribute irregularly, while an alternating copolymer possesses regularly alternating monomer residues. Block copolymers have two or more homopolymer subunits linked by covalent bonds, while graft copolymers contain side chains attached to the main polymer chains.

Polymer architecture: A polymer's architecture affects many of its physical properties, including viscosity, solubility, glass transition temperature, etc. A variety of techniques may be employed for the synthesis of a polymeric material with a range of architectures, such as star polymers, comb polymers, brush polymers, dendrimers, etc.

Chain conformation and configuration: Conformation and configuration are used to describe the geometric structure of a polymer chain, with conformation referring to order that arises from the rotation of molecules around the single bonds, while configuration concerns the order that is determined by the various chemical bonds in a polymer chain.

Intermolecular forces: The attractive forces between polymer chains play a large part in determining a polymer's properties. Because polymer chains are so long, these interchain forces are amplified far beyond the attractions between conventional molecules. The intermolecular forces in polymers can be affected by dipoles in the monomer units. Polymers containing amide or carbonyl groups can form hydrogen bonds between adjacent chains, where the partially positively charged hydrogen atoms in N—H groups of one chain are strongly attracted to the partially negatively charged oxygen atoms in C=O groups of another. These strong hydrogen bonds result in the high tensile strength and melting point of polymers containing urethane linkages, for example. Polyesters have dipole–dipole bonding between the oxygen atoms in C=O groups and the hydrogen atoms in H—C groups. Since dipole bonding is not as strong as hydrogen bonding, the melting point and strength of polyester are generally lower than those of polyamide, which has otherwise similar chemical composition. Ethene has no permanent dipoles and the attractive forces between polyethylene chains arise from weak van der Waals forces, which explains the low melting temperature of polyethylene.

Polymer morphology: Polymers may consist of both crystalline and amorphous regions, and polymer morphology describes the arrangement and microscale ordering of polymer chains in space. In the crystalline region, the polymeric chains are arranged in a three-dimensionally ordered state, usually arising from intramolecular folding and/or stacking of adjacent chains. The degree of crystallinity is expressed in terms of a weight fraction or volume fraction of the crystalline material, which ranges from zero for a completely amorphous polymer to one for a completely crystalline polymer. Crystalline polymers are generally tougher and more resistant to impact than totally amorphous polymers.

3.2.2 Polymer properties

Polymers are a unique type of material with novel chemical and physical properties. As a continuous macroscopic material, polymers derive their properties through the interactions of polymeric chains, either within the polymers on a microscale, or with other chemicals and solvents on a macroscale.

3.2.2.1 Solubility

The dissolution of polymeric materials is generally a slow process because polymers constitute long chains with a large number of segments, forming tightly folded coils which are entangled with each other. Due to their large size, coiled shape, and the attraction forces between them, polymer molecules dissolve much slower than low-molecular-weight molecules. There are two stages involved in the dissolution process: polymer swelling followed by dissolution. When a polymer is added to a given solvent, attraction and dispersion forces begin acting between its segments according to their polarity, chemical characteristics, and solubility parameter. If the polymer–solvent interactions are higher than the polymer–polymer attraction forces, the chain segments start to absorb solvent molecules, increasing the volume of the polymer matrix and loosening their coiled shape, leading eventually to the formation of a viscous polymer solution.

3.2.2.2 Mixing behavior

When a polymer is mixed with another polymer, an interpenetrating network is formed when the two polymers are structurally similar. Alternatively, polymers can precipitate each other to form a polymer blend with properties derived from the individual components. Because of their large molecular sizes, polymers are far less miscible than materials with smaller molecules. This effect results from the fact that the driving force for mixing is usually entropy, not interaction energy. In other words, miscible materials usually form a solution not because their interaction with each other is more favorable than their self-interaction, but because of an increase in entropy and hence free energy associated with increasing the amount of volume available to each component.

3.2.2.3 Phase behavior

When they are subjected to temperature change by either heating or cooling, polymers undergo phase changes between solid and fluid, crystalline and amorphous. The glass transition temperature (T_g) describes the temperature at which amorphous polymers undergo a transition from a brittle, glassy, amorphous solid to a rubbery, viscous, amorphous liquid. At a higher temperature, the melting point of a polymer refers to the temperature at which a crystalline or semicrystalline phase is transformed into a solid amorphous phase. Both the glass transition and melting temperatures may be engineered by altering the degree of branching or cross-linking in the polymer or by the addition of plasticizers.

3.2.2.4 Physical and mechanical properties

The physical and mechanical properties of polymeric materials are related to the intrinsic properties of the polymer itself, and are also dependent on the manufacturing conditions. Among the many parameters, the tensile strength of a polymeric material quantifies how much stress the material can endure before suffering permanent deformation. This is very important in applications that rely upon a polymer's physical strength or durability. For example, a rubber band with a higher tensile strength will hold a greater weight before snapping. Young's modulus quantifies the elasticity of the polymer, which is defined as the ratio of the stress along an axis over the strain along that axis. Similar to tensile strength, modulus is also an important property for a polymeric material—for example, high-performance fibers used for reinforcing plastics should have a high modulus to ensure a good result. Viscoelasticity, another key physical property of polymeric materials, is a property of materials that exhibit both viscous and elastic characteristics when undergoing deformation. It describes a complex time-dependent elastic response that exhibits hysteresis in the stress–strain curve when the load is removed.

3.2.3 Fiber-forming polymers

By forcing a viscous fluid through the tiny holes of a device called a spinneret, polymers can be converted into continuous filaments, which upon further processing can be made into textile fabrics in various forms. During the fiber-making process, the fiber-forming polymers are solids in their initial state and must be converted into a fluid state prior to extrusion. This is usually achieved by melting, if the polymers are thermoplastic, or by dissolving them in a suitable solvent if they cannot form a stable melt at high temperatures. If the polymers cannot be dissolved or melted directly, they can be chemically treated to form soluble or thermoplastic derivatives. Because of the requirements of the fiber-forming process and the requirements for the mechanical properties of the resultant fibers, fiber-forming polymers should have the following characteristics.

1. They should have a linear chain structure with no side chain or large side groups.
2. They should ideally have polar groups in the structure to give favorable performance regarding solubility, glass transition temperature, absorption, and other properties which are important for their application as a fibrous material.
3. They should have relatively flexible chain structures.
4. They should have regular chemical and stereo structures so that the resultant fibers can have an appropriate supramolecular structure.
5. They should have good thermal stability, with the softening temperature above the temperature of their routine application.
6. They should be stable in conventional solvents such as water, ethanol, etc.
7. The molecular weight and its distribution are also important for fiber-forming polymers. If the molecular weight is too high, the fiber-making process becomes difficult. On the other hand, if the molecular weight is too low, the resultant fibers would have insufficient mechanical properties.

Since Georges Audemars developed the first "artificial silk" in 1855 by dissolving mulberry bark in a solution and then dipping a needle into the solution to form a filament, a large number of chemical fibers have been developed over the years by using natural and synthetic polymers. Some of these are described below.

3.2.3.1 Natural fiber-forming polymers

Cellulose

Cellulose is the structural framework of plant cells, and is a linear polymer with repeating units consisting of D-glucose. It is the most abundant natural polymer and is used in a wide variety of applications. Due to the presence of strong hydrogen bonding between polymer chains, cellulose is insoluble in common solvents and many attempts have been made to dissolve it, which is a crucial step in the production of regenerated cellulose fibers. The viscose rayon process uses carbon disulfide to convert cellulose into cellulose xanthate, which is soluble in aqueous alkali solutions, before forming fibers in an acidic coagulation bath. In recent years a new solvent system comprising N-methylmorpholine N-oxide has been used to dissolve cellulose, which is a more environmentally friendly process for the production of cellulosic fibers.

Since the hydroxyl groups in cellulose are reactive and can be easily functionalized, cellulose has been made into a large number of derivatives in the form of ethers, esters, and acetals, such as methyl cellulose, hydroxypropylcellulose, hydroxypropyl methyl cellulose, and carboxymethyl cellulose. These cellulose derivatives are soluble in a variety of solvents and can be easily processed into various forms, such as membranes, sponges, and fibers. Due to their high diffusional permeability to most toxic metabolic solutes, cellulose membranes have been extensively investigated for use as hemodialysis membranes.

Chitin and chitosan

Chitin is a naturally occurring polysaccharide existing in the outer shells of crustaceans, insect exoskeletons, and fungal cell walls. It is the second most abundant natural polymer after cellulose. Chitin is commercially produced from the shell waste of crabs, shrimps, and krills through a series of deproteinization and demineralization processes to remove the protein and minerals, which together with chitin form the composite structure of the shells. The dry mass of shell waste typically contains about 15–25% of chitin.

Chemically, chitin is a (1→4) β-linked glycan composed of 2-acetamido-2-deoxy-D-glucose. Because of its highly crystalline structure, chitin is insoluble in common solvents. By hydrolyzing the acetamide groups in chitin into primary amine groups, the de-acetylated derivative chitosan can be easily dissolved in aqueous acidic solutions, which makes it suitable for the production of fibers. Chitosan undergoes biodegradation in vivo enzymatically by lysozyme to form nontoxic products. This easy processability and its versatile properties make chitosan an attractive material for various biomedical applications.

Proteins

Proteins are high-molecular-weight polymers in which the repeating amino acid monomers are joined together by peptide bonds. Since the major components of soft and hard tissues in the human body are composed of proteins, these polymers have been extensively investigated for various applications, such as sutures, hemostatic agents, scaffolds for tissue engineering, and drug delivery.

Collagen is the most abundant protein in the human body, serving as the major component of bone, skin, ligament, cartilage, and tendon. Its basic component is a polypeptide comprising the repeating sequence of glycine, proline, and hydroxyproline. This polypeptide combines with 12 others to form the left-handed triple-helix structure in collagen. Due to its enzymatic degradability and unique physicochemical, mechanical, and biological properties, collagen has been extensively investigated for various biomedical applications. It can be processed into different forms, such as fibers, sheets, tubes, sponges, powders, fleeces, injectable solutions, and dispersions. The use of collagen as a suture material dates back for a millennium, and one form of it, catgut, is still widely used today.

Gelatin is prepared by the thermal denaturation of collagen, usually isolated from animal skin and bones in the presence of dilute acid. Structurally, gelatin consists of 19 amino acids joined by peptide linkages, and its enzymatic degradation results in the formation of the corresponding amino acids. Biologically, gelatin does not show antigenicity and has good hemostatic properties.

Alginic acid

Alginic acid is a linear heteropolysaccharide obtained from the cell walls of brown algae. It is composed of D-mannuronic acid and L-guluronic acid in different proportions depending on the types of seaweed. Due to the presence of carboxylic acid groups along the polymer chain, water-soluble alginate salts such as sodium alginate can form gel in the presence of divalent metal ions such as calcium ions. Gelation occurs by the interaction of divalent cations with blocks of guluronic acid from different polysaccharide chains, resulting in the so-called "egg-box" structure. Because of their ability to hold a large amount of water, alginate-based materials such as fibers, foams, and hydrogels have been widely used in cosmetic and biomedical applications.

Hyaluronic acid

Hyaluronic acid is a naturally occurring linear anionic polysaccharide consisting of repeating disaccharide units. It forms an important component of articular cartilage and is widely distributed in the connective tissues of mammals. Chemically, hyaluronic acid consists of D-glucuronic acid and 2-acetamido-2-deoxy-D-glucose monosaccharide units. The polymer is water soluble and forms very viscous solutions. Hyaluronic acid possesses several properties that make it an ideal candidate for wound-dressing applications. For example, it can act as a scavenger for free radicals in wound sites, thereby modulating inflammation. It can interact with a variety of biomolecules and act as a bacteriostat. More importantly, it can be recognized by receptors on a variety of cells associated with tissue repair.

3.2.3.2 Synthetic fiber-forming polymers

Polyethylene terephthalate

Polyethylene terephthalate (PET) is a thermoplastic polymer of the polyester family and is widely used in the production of synthetic fibers. PET was patented in 1941 by John Rex Whinfield, James Tennant Dickson, and their employer, the Calico Printers' Association of Manchester. During the polymerization process, ethylene glycol and dimethyl terephthalate ($C_6H_4(CO_2CH_3)_2$) or terephthalic acid are reacted in the melt state at 150–200 °C with a basic catalyst. As part of the production process, methanol is removed by distillation to drive the reaction forward, and excess ethylene glycol is distilled off at a higher temperature with the aid of vacuum. The second transesterification step proceeds at 270–280 °C, with continuous distillation of ethylene glycol as well. The reactions are idealized as follows:

First step:

$$C_6H_4(CO_2CH_3)_2 + 2\,HOCH_2CH_2OH \rightarrow C_6H_4(CO_2CH_2CH_2OH)_2 + 2\,CH_3OH$$

Second step:

$$n\,C_6H_4(CO_2CH_2CH_2OH)_2 \rightarrow [(CO)C_6H_4(CO_2CH_2CH_2O)]_n + n\,HOCH_2CH_2OH$$

In the terephthalic acid process, esterification of ethylene glycol and terephthalic acid is conducted directly at a moderate pressure (2.7–5.5 bar) and a high temperature (220–260 °C). Water is eliminated in the reaction, and is also continuously removed by distillation:

$$n\,C_6H_4(CO_2H)_2 + n\,HOCH_2CH_2OH \rightarrow [(CO)C_6H_4(CO_2CH_2CH_2O)]_n + 2n\,H_2O$$

Polyamide

In the polyamide (PA) polymers, the amide link is produced from the condensation reaction of an amino group and a carboxylic acid or acid chloride group. A small molecule, usually water or hydrogen chloride, is eliminated. The amino group and the carboxylic acid group can be on the same monomer, or the polymer can be constituted of two different bifunctional monomers, one with two amino groups and the other with two carboxylic acid or acid chloride groups. The two main types of polyamides used in the textile industry are PA6 and PA66: PA6 $(NH—(CH_2)_5—CO)_n$ is made from ε-Caprolactam, and PA66 $(NH—(CH_2)_6—NH—CO—(CH_2)_4—CO)_n$ is made from hexamethylenediamine and adipic acid.

Polyacrylonitrile

Polyacrylonitrile (PAN) is a synthetic polymer with the linear formula $(C_3H_3N)_n$. Although it is thermoplastic it does not melt under normal conditions, since it degrades before melting at above 300 °C. Almost all PAN resins are copolymers made from mixtures of monomers, with acrylonitrile as the main component. It is a versatile polymer used to produce a large variety of products, including fibers for textiles, ultra-filtration membranes, hollow fibers for reverse osmosis, etc. PAN fibers are the precursor in the production of high-quality carbon fibers.

Polypropylene

Polypropylene (PP) is a thermoplastic polymer used in a wide variety of applications, including textile fibers, packaging and labeling, plastic parts and reusable containers of various types, laboratory equipment, automotive components, etc. PP is normally tough and flexible, especially when copolymerized with ethylene. It has good resistance to fatigue, and is unusually resistant to many chemical solvents such as bases and acids.

Polyethylene

Polyethylene (PE) is the most common plastic, with annual production at approximately 80 million metric tons. Its primary use is within packaging, such as plastic bags, plastic films, geomembranes, containers, bottles, etc. Depending on its stereo regularity, PE is divided into low-, medium-, and high-density categories, with melting points in the range of 105–130 °C. PE with an extremely high molecular weight can be made into the high-strength fibers used in bullet-proof vests.

Polyurethanes

Polyurethane (PU) is a polymer composed of a chain of organic units joined by carbamate (urethane) links. PU polymers are formed by combining two bi-functional (or higher) monomers, one of which contains two or more isocyanate functional groups with the formula —N=C=O, while the other contains two or more hydroxyl groups. The isocyanate and hydroxyl groups combine to form a urethane linkage:

$$ROH + R'NCO \rightarrow ROC(O)N(H)R' \text{ (R and R' are alkyl or aryl groups)}$$

PUs are widely used in the manufacture of flexible foam materials. In the textile industry, these polymers have been made into fibers with excellent elasticity. Spandex, which is a PU-based elastic fiber, has been incorporated into a wide range of textile fabrics, including elastic stockings for compression therapy.

Polyvinyl chloride

Polyvinyl chloride (PVC) is the third most widely produced plastic, after PE and PP. Pure PVC is a white, brittle solid which is insoluble in alcohol but slightly soluble in tetrahydrofuran. PVC can be made softer and more flexible by the addition of plasticizers, the most widely used being phthalates. In this form it is used in clothing and upholstery, electrical cable insulation, inflatable products, and many applications in which it replaces rubber.

Polyvinyl alcohol

Polyvinyl alcohol (PVA) is a water-soluble synthetic polymer with excellent film-forming, emulsifying, and adhesive properties. It has high tensile strength and flexibility, as well as high oxygen and aroma barrier properties. PVA can be made into fibers by either wet spinning or dry spinning. Its water-soluble nature makes it a good raw material for many biomedical products, such as hydrogels and wound dressings.

Polytetrafluoroethylene

Polytetrafluoroethylene (PTFE), a synthetic fluoropolymer of tetrafluoroethylene, is a high-molecular-weight compound consisting wholly of carbon and fluorine. It is hydrophobic and has one of the lowest coefficients of friction against any solid. Because of the high strength of its carbon–fluorine bonds, PTFE is extremely stable and can be used for coating medical devices to ensure product stability.

Aramid

Aramid, or aromatic polyamides, were first introduced in commercial applications in the early 1960s, with a meta-aramid fiber produced by DuPont under the trade name Nomex. This fiber, which handles similarly to normal textile apparel fibers, is characterized by its excellent resistance to heat, as it neither melts nor ignites in normal levels of oxygen. It is used extensively in the production of protective apparel, air filtration, and thermal and electrical insulation. A para-aramid fiber with much higher tenacity and elastic modulus was also developed in the 1960s–1970s by DuPont, which was the first company to introduce a para-aramid fiber, Kevlar, in 1973. A similar fiber called Twaron with roughly the same chemical structure was introduced by Akzo in 1978.

Aliphatic polyesters

Aliphatic polyesters can be considered as representatives of synthetic biodegradable polymers. Synthesis of aliphatic polyesters by polycondensation of diols and dicarboxylic acids was reported as early as 1930. However, the low melting points, low hydrolytic stability, and low molecular weights of the polymers initially obtained severely limited their application. At the same time, their high hydrolytic instability resulted in a multitude of applications for this class of polymers in the biomedical field, starting with absorbable sutures in the 1960s. This new avenue revitalized interest in these polymers, and novel synthetic methods, as well as catalytic systems, were developed to obtain high-molecular-weight polymers with narrow molecular weight distributions.

The commonly used monomers for the synthesis of aliphatic polyesters for biomedical applications are lactide, glycolide, and caprolactone. Polyglycolic acid (PGA) was one of the initially investigated biodegradable polyesters for biomedical applications. It is a highly crystalline polymer with a melting point greater than 200 °C and a glass transition temperature around 35–40 °C. PGA was used to develop the first biodegradable synthetic suture, DEXON, in 1970.

Similar to PGA, polymers obtained from the synthesis of lactide have also been widely used in the biomedical field. As lactide is a chiral molecule, polymeric lactide exists as three isomers: L-lactide, D-lactide, and meso-lactide. Therefore, four different types of polylactic acid are available: poly(L-lactic acid), poly(D-lactic acid), poly(DL-lactic acid), and mesopolylactic acid.

Polycaprolactone (PCL) is a semicrystalline polyester with a melting temperature of about 55–60 °C. PCL is of great interest as it can be obtained from cheap starting material and has high solubility in organic solvents, low melting point, and an exceptional ability to form blends with a variety of polymers. Due to the low degradation rate of PCL, it has been investigated as a material for the long-term controlled delivery of drugs.

Polyanhydrides

Polyanhydrides form a class of surface-eroding polymers that have been extensively studied solely for biomedical applications. Although the first synthesis of polyanhydrides was reported as early as 1909, the low hydrolytic stability of aliphatic anhydrides coupled with the low molecular weight of many of these polymers barred them from any industrial application until 1980, when they were suggested as ideal candidates for drug-delivery applications. Various polymerization techniques have been developed for the synthesis of very-high-molecular-weight polymers.

Polyamino acids

Synthetic polyamino acids have been investigated for various biomedical applications due to their structural similarity with naturally occurring proteins. Although several homo- and co-polyamino acids have been synthesized and evaluated, their high crystallinity, low degradation rate, unfavorable mechanical properties, and immunogenicity severely limited their biomedical applications. Alternatively, amino acids can be combined with polymeric materials by grafting them on to polymer chains, copolymerizing with other monomers, synthesizing block copolymers with amino acid sequences, etc.

3.3 Fiber classification

The classification of fibers has been defined by many people, and it is generally accepted that they can be divided broadly into natural fibers and chemical fibers. Figure 3.2 shows the classification of textile fibers.

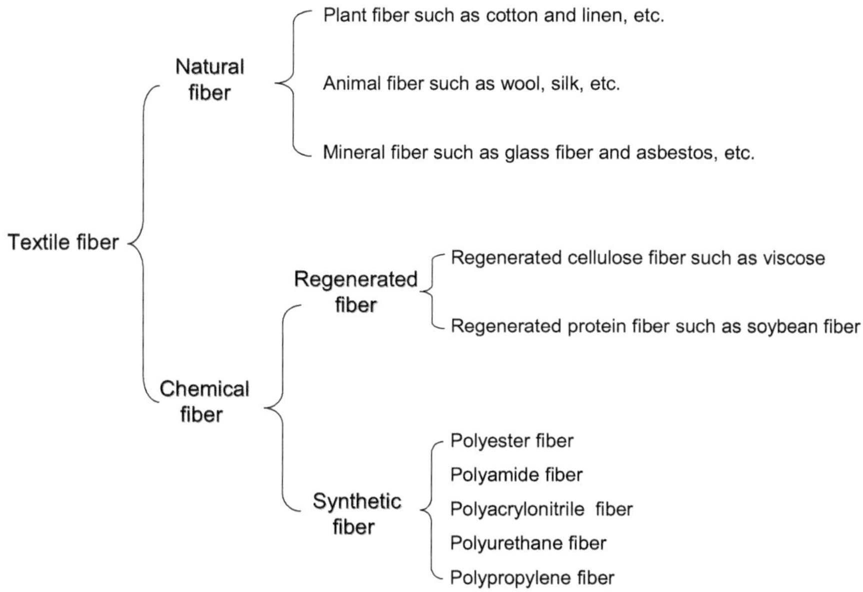

Figure 3.2 Classification of textile fibers.

3.3.1 Natural fibers

Natural fibers are those produced by plants, animals, and geological processes. They are biodegradable over time, and can be classified according to their origin into plant fibers, animal fibers, and mineral fibers.

3.3.1.1 Plant fibers

Plant fibers are generally composed of cellulose, often in combination with other components such as lignin. Examples include cotton, hemp, jute, flax, ramie, sisal, and bagasse. Plant fibers are classified according to their source in plants.

- Bast or stem fibers, derived from the fibrous bundles in the inner bark of plant stems.
- Leaf fibers, which run lengthwise through the leaves of monocotyledonous plants.
- Seed-hair fibers, such as cotton.

Cotton is a soft fiber that grows around the seeds of the cotton plant. Cool, soft, and comfortable, cotton is presently the world's most-used fiber, with applications in apparel, home furnishings, medical and surgical materials, and many other industrial products. Cotton fabrics have a pleasant luster, a soft drape, and a smooth hand. They are very comfortable to wear due to their soft hand and other characteristics. Cotton fabrics have excellent absorbing capabilities and, by absorbing perspiration, they keep the wearer more comfortable. Absorbent cotton products can retain 24–27 times their own weight of water and are stronger when wet than dry.

3.3.1.2 Animal fibers

Animal fibers consist largely of proteins. Examples are silkworm silk, spider silk, sinew, catgut, wool, animal hairs such as cashmere, mohair, and rabbit hair, etc.

3.3.1.3 Mineral fibers

Asbestos is the only naturally occurring long mineral fiber. It has many varieties, such as chrysotile, amosite, crocidolite, tremolite, anthophyllite, and actinolite.

3.3.2 Chemical fibers

Chemical fibers are fibrous materials produced from organic and inorganic raw materials in a chemical process. The organic materials may be natural or synthetic polymers, while the inorganic compounds include glass, metal, basalt, quartz, and other composites. Chemical fibers are manufactured industrially in the form of monofilaments, staple fibers, and filament yarns.

3.3.2.1 Fibers from natural polymers

Natural polymers can be made into fibers through dissolution of the polymer in an appropriate solvent and then extrusion of the polymer solution into a coagulation bath. As an example, cellulose can be made into viscose rayon fibers, cuprammonium rayon, cellulose acetate and triacetate fibers, lyocell, and modal fibers depending on the processes used to make the fibers. Other natural polymers such as rubber, chitosan, alginic acid, and protein can also be made into fibers in an appropriate fiber-forming process.

3.3.2.2 Fibers from synthetic polymers

There are many types of fibers made from synthetic polymers. The most common are polyester, PA, acrylic and modacrylic, PP, PUs, etc.

3.3.2.3 Fibers from inorganic materials

Inorganic chemical fibers are made from materials such as glass, metal, carbon, or ceramic. These fibers are often used to reinforce plastics to form composites.

3.4 Fiber structure

Fiber structure is generally known in micro and macro terms. On the micro level, a textile fiber is made of polymeric chains that are organized into three distinct phases: the oriented crystalline regions; the amorphous regions, also with preferential orientation along the fiber axis which contain tie molecules connecting crystallites; and the highly extended noncrystalline molecules, which are called the interfibrillar phase. The degree of orientation and degree of crystallinity are two important structural characteristics of a textile fiber. The first describes how well the polymer chains within a fiber align along the fiber axis, while the second is concerned with the three-dimensional order of the polymer chains. Both are influenced by manufacturing conditions, and to some extent can be controlled.

On the macro level, the main structural characteristics of a textile fiber are its surface structure and cross-sectional size and shape. Different types of fibers differ greatly on the macro level. For example, polyester and nylon fibers usually have smooth surface and round cross-sectional structures, while natural fibers such as cotton and wool have rough surface structures and irregular cross-sectional shapes.

3.5 Fiber properties

As a special form of polymeric material, fibers possess many unique properties. They are characterized by their flexibility, fineness, and a high ratio of length to width. The extreme fineness is shown by the fact that 1 kg of cotton fibers would contain about 10^9 individual fibers. Fibers can be utilized in a number of special applications because they are known to have the following properties.

1. Combination of flexibility and strength.
2. Crack stopping.
3. Continuity over long lengths.
4. High surface area and entrapped space.
5. Versatile processibility into many forms of assemblies.

Technically, fibers are characterized by their physical, mechanical, chemical, and biological properties.

3.5.1 Physical and mechanical properties of textile fibers

The physical and mechanical properties of textile fibers include their morphological appearance, density, strength, extensibility, thermal and electrical conductivity, and many others. On the morphological level, the aspect ratio, that is, the ratio of length to diameter of a fiber, is an important characteristic, while luster is also a key property of textile fibers. Luster defines the quality of shining with reflected light, which is frequently associated with the adjectives "bright" or "dull" to distinguish between varieties of manufactured fibers.

The aesthetic appearance of a textile fiber relates to properties perceived by touch and sight, such as the hand, color, luster, drape, and texture of fabrics or garments. The waviness of a fiber is expressed as crimps per unit length, which measures the difference in distance between two points on an unstretched fiber and the same two points when the fiber is straightened under specified tension.

Density is defined as the mass per unit volume, usually expressed as grams per cubic centimeter. Because textile fibers are made from a large variety of materials, they differ greatly in their density. For example, cotton has a relatively high density at $1.54\,g/cm^3$, while polyester fibers have a density of $1.38\,g/cm^3$.

Linear density refers to the mass per unit length expressed as grams per 9000 or 10,000 m of fiber or yarn, which are known respectively as denier or decitex. Denier and decitex are direct numbering systems in which the lower numbers represent the finer sizes and the higher numbers the coarser sizes. In the US the denier system is used for numbering filament yarns, staple fibers, and tow. In most countries outside the US the denier system has been replaced by the decitex system or the tex system. Tex is also a unit for expressing linear density, equal to the weight in grams of 1 km of yarn, filament, fiber, or other textile strand.

The electrical property of a textile fiber is mainly concerned with the accumulation of negative or positive electricity on the surface of fibers or fabrics because of inadequate electrical dissipation during processing. Static charge results in an electrical attraction or repulsion of the fibers relative to themselves, to machine parts, or to other materials, preventing the fiber from traveling in a normal path in the process. Antistatic properties refer to the ability of a textile material to disperse an electrostatic charge and prevent the buildup of static electricity.

In addition to electrical charge, textile fibers are influenced by external conditions such as oxidation and flame. Deterioration of textiles or other materials can be caused by gradual oxidation during storage and/or exposure to light. Flame resistance is a term used to describe a material that burns slowly or is self-extinguishing after the removal of an external source of ignition. Fibrous materials can be flame resistant because of the innate properties of the fiber, the twist level of the yarn, the fabric construction, the presence of flame retardants, or a combination of these factors.

Dimensional stability is another important physical property of textile fibers. Shrinkage describes the widthwise or lengthwise contraction of a fiber, yarn, or fabric, usually after wetting and redrying or on exposure to elevated temperatures.

The mechanical properties of textile fibers describe the characteristics of a fiber when it is subjected to some type of force that attempts to stretch, twist, or fracture it. Many tests have been developed to quantify these mechanical properties. Tensile test is a method of measuring the resistance of a fiber, yarn, or piece of fabric to a force tending to stretch the specimen in one direction. Tensile stress refers to the resistance to deformation developed within a specimen subjected to tension by external force. It is commonly expressed in two ways, either as tensile strength, that is, the force per unit cross-sectional area of the unstrained specimen, or as tenacity, that is, the force per unit linear density of the unstrained specimen. The latter is more frequently used in textile testing. Tensile strain is the relative length of deformation exhibited by a specimen subjected to a tensile force. For a textile fiber, elongation at break defines the strain as a percentage of the original length at the point when the fiber fractures under tensile stress. Young's modulus, a property of a perfectly elastic material, is the ratio of change in stress to change in strain within the elastic limits of the material. The ratio is calculated from the stress expressed in force per unit cross-sectional area and the strain expressed as a fraction of the original length. Modulus so calculated is equivalent to the force required to strain the sample 100% of its original length, at the rate prevailing below the elastic limit.

Figure 3.3 shows the stress and strain curves of some typical textile fibers, and the relationship between the change in dimension in the direction of the applied stress

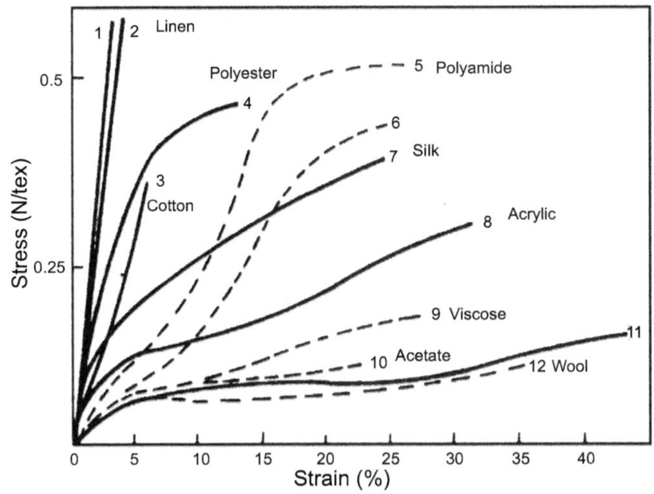

Figure 3.3 Stress and strain curves of some typical textile fibers.

and the magnitude of that stress. In tension tests of textile materials the stress can be expressed either in units of force per unit cross-sectional area or in force per unit linear density of the original specimen, and the strain can be expressed as either a fraction or a percentage of the original length. Sometimes the stress–strain curve is also expressed as the load-extension curve.

3.5.2 Chemical properties of textile fibers

Textile fibers are organic materials with large surface areas. Because of the various reactive chemical groups, fibers can be chemically modified by using a number of processes. In particular, fibers have absorption properties for gases, liquids, or dye molecules.

The ability of a textile fiber to absorb water from the atmosphere is important in many applications. In addition to the chemical and physical structures of the fiber, the temperature and water content of the atmosphere strongly influence the amount of water that is absorbed by the fiber. In this respect, a standard atmosphere is defined as air maintained at 20 °C and 65% relative humidity. Special humidity and temperature conditions are sometimes prescribed for the testing of textiles for specific purposes. The test specimens are often brought to moisture equilibrium in the standard atmosphere before tests are carried out.

All fibers can absorb some moisture when exposed to the atmosphere. The quantity varies with the fiber type, temperature, and relative humidity. There are two ways to characterize the absorption properties of fibers. First, moisture content measures the amount of water absorbed by a fiber or yarn as a percentage of the total weight of the preconditioned fiber, while moisture regain is defined as the percentage of moisture in a textile material brought into equilibrium with a standard atmosphere after drying, calculated as a percentage of the moisture-free weight. Table 3.1 shows the moisture regain and other main properties of some textile fibers.

Table 3.1 Main properties of textile fibers

Type of fiber	Density (g/cm^3)	Moisture regain (%)	Tenacity (g/dtex)	Elongation at break (%)
Cotton	1.54	7–8.5	2.3–4.5	3–10
Wool	1.32	14–16	0.9–1.8	30–45
Viscose rayon	1.52	12–16	1.5–4.5	9–36
Cellulose acetate	1.30	6–6.5	1.0–1.26	23–45
Polyacrylonitrile	1.17	1.5	1.8–4.5	16–50
Polyester	1.38	0.4	2.5–5.5	10–45
Polyamide	1.14	4–4.5	3.6–8	16–45
Alginate	1.78	17–23	0.9–1.8	2–14
Chitosan	1.39	10–12.5	1.2–2.2	7–33

The ability to absorb dye molecules is another important chemical property of a fiber. Once dyed, fibers should have good color fastness and be able to resist fading, which relate to the property of a dye to retain its color when the dyed or printed textile material is exposed to conditions or agents such as light, perspiration, atmospheric gases, or washing that can remove or destroy the color.

3.5.3 Biological properties of textile fibers

The biological properties of textile fibers describe their response under the influence of microorganisms. Antimicrobial fibers are often used in medical applications, socks, shoe liners, etc.; antimicrobial agents are introduced directly into the fiber matrix as opposed to fiber having an antimicrobial finish applied. Biodegradable fibers are those fibers that are able to be broken down by microorganisms so that they can be returned to the environment without causing pollution.

3.6 Methods for the manufacture of chemical fibers

Nature is full of fibers and fibrous materials, and it has been consistently recognized that the science and technology of fibers were learnt from silkworms and spiders, since their fiber-producing processes are good examples of biosynthetic and biospinning techniques in which they convert nonfiber foods by enzymes into proteins in the body and then spin fibers as a cocoon or net.

The first attempt to create an artificial fiber was made in 1664 by an English naturalist, Robert Hooke, who suggested the possibility of producing a fiber. In 1855, the first patent for "artificial silk" was granted in England to a Swiss chemist, Georges Audemars. By dissolving the fibrous inner bark of a mulberry tree and chemically modifying it, Audemars was able to produce cellulose fibers by extruding the cellulosic liquid through a small hole.

The first commercial-scale production of a manufactured fiber was achieved by a French chemist, Count Hilaire de Chardonnet. In 1884 his fabrics made of "artificial silk" caused a sensation at the Paris Exhibition. Two years later he built the first commercial rayon plant at Besançon, France, and secured his fame as the "father of the rayon industry." However, the artificial silk invented by Chardonnet was based on cellulose nitrate, and its flammability limited its applications. In 1892 the viscose rayon process was invented by Charles F. Cross, Edward J. Bevan, and Clayton Beadle in England, and became the basis for the regenerated cellulose fiber industry.

In September 1931 American chemist Wallace Carothers reported on research carried out in the laboratories of the DuPont Company on "giant" molecules called polymers. He focused his work on a fiber referred to simply as "66," a number derived from its molecular structure. Nylon, the "miracle fiber," was born. By 1938 Paul Schlack of the I.G. Farben Company in Germany had polymerized caprolactam and created a different form of the polymer, identified simply as nylon "6." Meanwhile polyester, first examined as part of Wallace Carothers's early research, was attracting new interest at the Calico Printers' Association in Great Britain, where John Whinfield and James Dickson produced a polyester fiber by condensation polymerization of ethylene glycol with terephthalic acid. DuPont subsequently acquired the US patent rights and Imperial Chemical Industries (ICI) the rights for the rest of the world. Polyester's commercialization in 1953 was accompanied by the introduction of cellulose triacetate. The majority of the twentieth century's basic manufactured fibers had now been discovered, and the industry's engineers turned to refining their chemical and physical properties. In the 1960s manufactured fiber production accelerated, spurred on by continuous innovation. The revolutionary new fibers were modified to offer greater comfort, provide flame resistance, and have easier dyeability and better blending qualities.

Technically, chemical fibers are commonly made by extruding a thick, viscous liquid through a spinneret to form continuous filaments. In this process, the fiber-forming polymers are initially in a solid state and must be converted into a fluid state prior to extrusion. This is usually achieved by melting, if the polymers are thermoplastic, or by dissolving them in a suitable solvent if they are nonthermoplastic. If they cannot be dissolved or melted directly, they must be chemically treated to form soluble or thermoplastic derivatives.

Figure 3.4 shows the formation of fibers during the melt and wet spinning processes. As the filaments emerge from the holes in the spinneret, the liquid polymer is first converted to a rubbery state and then solidified. This process of extrusion and solidification of endless filaments is called spinning, and shares the same name as the textile operation whereby short pieces of staple fibers are twisted into yarn. There are four basic methods of spinning filaments of manufactured fibers: melt, wet, dry, and gel spinning. Other relatively new techniques, such as electrospinning, have also been used to make chemical fibers.

3.6.1 Melt spinning

Melt spinning is a manufacturing process used to produce synthetic fibers. This method can be used to create fibers for fabrics or to spin glass into fine threads for fiberglass

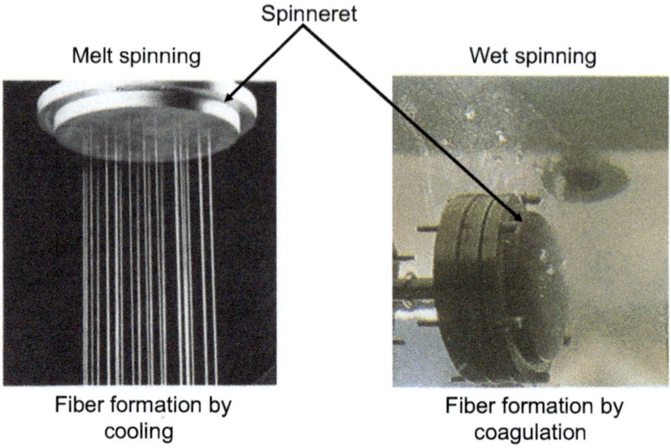

Figure 3.4 The formation of fibers during the melt and wet spinning processes.

production. It utilizes a rapid cooling system to transform melted base materials into long strands or filaments. Melt spinning is commonly used to produce fibers such as nylon, polyester, PP, etc.

In a typical melt spinning process, the manufacturing operation is performed on a large spinning tower. A drum of melted polymers sits at the top of the tower, with a metering pump and filtering system situated below. The pump directs melted polymer down through a metal die, or spinneret. This die contains a large number of microscopic holes that help to form the molten polymer into thin strands.

In this process, as the filaments exit the spinneret they immediately pass in front of a cold air blower, which rapidly cools and solidifies the strands. Then they travel down to a series of advancing rollers, which guide the filaments through various processes, such as stretching, crimping, heat setting, etc., before they are wound up as rolls of continuous filaments or cut into short staple fibers.

3.6.2 Solution wet spinning

In the wet spinning process, the spinning solution is initially pumped by a precision gear pump through a spinneret into a coagulation bath, which precipitates the polymer to form filaments. At the exit of the coagulation bath, filaments are collected in bundles of the desired thickness. The multifilament bundles are counter-currently washed in successive extraction baths to remove residual solvent before they are stretched, dried, and heat-relaxed.

3.6.3 Solution dry spinning

In the dry spinning method, the polymer solution is heated before being extruded into a hot stream of gas, which is also continuously heated during its passage. Fibers are formed by evaporating the solvent. In this process, the fiber surface hardens initially while the solvent still diffuses out from the interior of the filament, and a cylinder under external overpressure is formed. The internal pressure decreases with increasing distance from the spinneret until the surface suddenly collapses to form a round, oval, or serrated, deformed cross-section depending on the nature of the polymer.

3.6.4 Dry-jet wet spinning

In the dry-jet wet spinning process, the polymer solution is extruded from a dry jet and subsequently followed by coagulation as in conventional wet spinning. The polymer solution enters into an air gap before it is precipitated in the coagulation bath, hence the name dry-jet wet spinning. This method can allow stress relaxation for polymer chains in the air gap, and as a result the as-spun fibers are less oriented and more uniform than those from an immersed jet. This permits orientation by subsequent drawing to produce fibers with higher tenacity.

3.6.5 Gel spinning

Gel spinning is a special process used to obtain high-strength fibers. In this process, the fiber-forming polymer is not in a true liquid state during extrusion. Not completely separated, as they would be in a true solution, the polymer chains are bound together at various points in liquid crystal form. This produces strong interchain forces in the resulting filaments, which can significantly increase the tensile strength of the fibers. In addition, the liquid crystals are aligned along the fiber axis by the shear forces during extrusion, which makes it possible to obtain filaments with an unusually high degree of orientation. This process can also be described as dry-wet spinning, since the filaments first pass through air and are then cooled further in a liquid bath. High-strength PE and aramid fibers are produced by gel spinning.

3.6.6 Phase-separation spinning

Phase-separation spinning is concerned with the extrusion of polymers and solvent at high temperatures into a cooling zone. During the cooling process a phase separation occurs, usually accompanied by crystallization of the solvent. The polymers in the extruded filament form fibers during this process, while the solvent can be removed before or after drawing.

3.6.7 Reaction spinning

Reaction spinning is a process in which an initial prepolymer is formed and then extruded into a reagent bath, where polymerization and filament formation occur simultaneously. PU-based elastic fibers can be made by this method.

3.6.8 Electrospinning

Electrospinning is a novel fiber-making process in which polymers from a solution or melt can be deposited as fibrous material by charging the liquid to about 5–30 kV and ejecting it through a nozzle on to an oppositely charged grounded target. An electrospinning system consists of a high-voltage DC supply, a grounded electrode, a nozzle system with diameter controls, and a fixed or rotated target to which the spun fiber can be adhered. This spinning method can be designed to offer fibers with excellent surface morphology and porosity to provide the most appropriate interface for biomedical applications. The fibers are often spun on to nonwoven structures which are porous and have a high surface area, with the resulting electrospun nonwoven structure ideally suited for making scaffolding for tissue engineering.

3.7 Summary

Textile fibers are a unique form of polymeric material which combine the chemical, physical, and biological properties of polymers and the flexibility, strength, and processibility of fibrous structures. While natural fibers such as cotton, linen, and silk have been used in the medical textile industry for a long time, chemical fibers can be made from a diversified range of polymers in a variety of ways, which makes it possible to engineer the performance of fibers according to end-use requirements. In this sense, many fiber-making techniques such as solution spinning, melt spinning, and more importantly electrospinning can be utilized to engineer fibrous structures that generate functional benefits for medical textile materials.

Sources for further information

[1] Cook JG. Handbook of textile fibers. Cambridge: Woodhead Publishing Ltd; 2001.
[2] Hearle JWS. High-performance fibers. Cambridge: Woodhead Publishing Ltd; 2001.
[3] Hongu T, Phillips GO, Takigami M. New millennium fibers. Cambridge: Woodhead Publishing Ltd; 2005.
[4] Beyreuther R, Brunig H. Dynamics of fiber formation and processing: modelling and application in fiber and textile industry. Berlin: Springer; 2007.
[5] Walczak ZK. Processes of fiber formation. London: Elsevier Science Ltd; 2002.
[6] Williams DF. The williams dictionary of biomaterials. Liverpool: Liverpool University Press; 1999.
[7] Barbucci R, editor. Integrated biomaterial science. New York: Kluwer/Plenum Press; 2002.
[8] Rajendran S, editor. Advanced textiles for wound care. Cambridge: Woodhead Publishing Ltd; 2009.
[9] Long AC, editor. Design and manufacture of textile composites. Cambridge: Woodhead Publishing Ltd; 2005.
[10] Taylor MA. Technology of textile properties. London: Forbes Publications Ltd; 1990.

A brief description of the manufacturing processes for medical textile materials

4

4.1 Introduction

Textile materials are formed by weaving, knitting, braiding, crocheting, knotting, or pressing fibers together using a number of different methods, such that individual fibers are assembled to produce a two- or three-dimensional structure with sufficient mechanical properties. In the textile industry, the word "textile" refers to any material made of interlacing fibers, while the word "fabric" is used to describe any material made by weaving, knitting, crocheting, or bonding yarns or fibers. Another word, "cloth," may be used synonymously with "fabric," but often refers to a finished piece of fabric used for a specific purpose, such as a tablecloth.

Traditionally, fibers are first spun into yarn or thread before they are made into fabric. A number of production methods can be used. For example, weaving is a textile production method which involves interlacing a set of longer threads (called the warp) with a set of crossing threads (called the weft) on a frame or machine known as a loom. Knitting and crocheting involve interlacing loops of yarn, which are formed either on a knitting needle or on a crochet hook, together in a line. Braiding or plaiting involves twisting threads together into cloth, while knotting involves tying threads together and is used in making macrame. Lace is made by interlocking threads together independently, using a backing and any of the methods described above, to create a fine fabric with open holes in the work. Carpets, velvet, velour, and velveteen are made by interlacing a secondary yarn through woven cloth to create a tufted layer. Felting involves pressing a mat of fibers and working them together until they become tangled. A liquid, such as soapy water, is usually added to lubricate the fibers and open up the microscopic scales on strands of wool. Non-woven textiles are manufactured by the bonding of fibers to make fabric: the bonding may be thermal, mechanical, or use adhesives. Figure 4.1 shows some of the commonly used textile structures.

In order to possess the right characteristics to meet end-use requirement, fibers, yarns, and textile fabrics are often treated in a number of chemical, physical, and biological processes, such as bleaching to remove impurities and dyeing to give the material appropriate color and appearance. In addition to these traditional processes, textile materials used for medical applications need to be packaged and sterilized before they can meet the functional and safety requirements of a medical product.

Medical Textile Materials. http://dx.doi.org/10.1016/B978-0-08-100618-4.00004-2

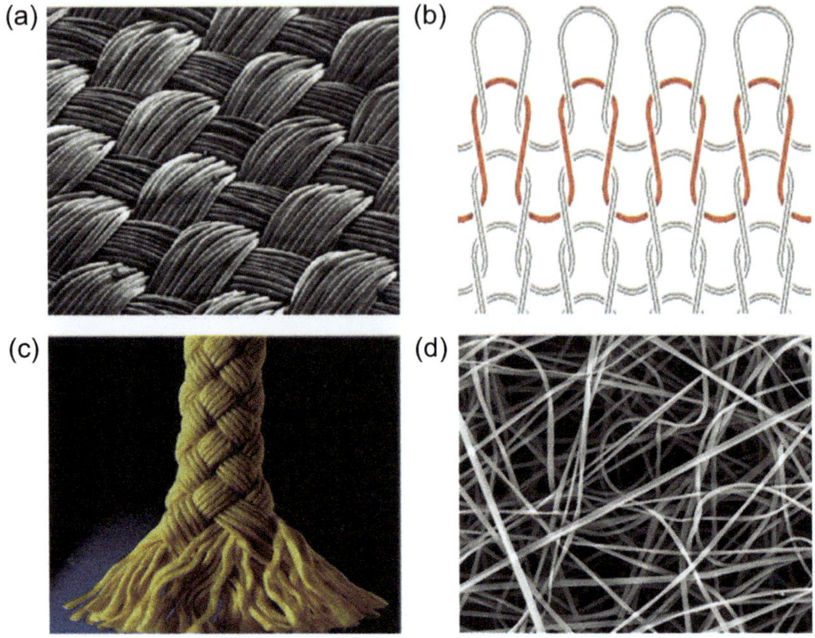

Figure 4.1 An illustration of the commonly used textile structures: (a) woven; (b) knitted; (c) braided; (d) nonwoven.

4.2 Yarn processing

Yarn is a long continuous length of interlocked fibers, suitable for use in sewing, crocheting, knitting, weaving, embroidery, and ropemaking. Yarns are made by utilizing either staple or filament fibers, or by combining both. Filament yarn is a term applied to a long rope-like bundle of raw fibers which is extruded from the spinneret in the chemical fiber production process. Filament yarns can be monofilament or multifilament. Spun yarn is made by gathering a bundle of staple fibers and spinning the spindles at a very high speed to twist the staple fibers together to form a piece of yarn.

In the textile industry, yarn spinning was one of the very first processes to be industrialized. Spun yarns may contain a single type of fiber or a blend of various types. Combining synthetic fibers (which can have high strength, luster, and fire-retardant qualities) with natural fibers (which have good water absorbency and skin-comforting qualities) is very common. The most widely used blends are cotton-polyester and wool-acrylic fibers. Blends of different natural fibers are common too, especially with more expensive fibers such as cashmere. Yarns are selected for different textile products based on the characteristics of the constituent fibers, such as wool for warmth, nylon for durability, cashmere for softness, etc.

Technically, yarns are made up of a number of singles, which are known as plies when grouped together. These singles of yarn are twisted together in the opposite direction to make a thicker yarn. Depending on the direction of the final twist, the yarn can be S-twist or Z-twist.

Filament yarn consists of filament fibers either twisted together or simply grouped together. In order to develop stretch and bulk in subsequent processing, filament yarns are often texturized. When woven or knitted into fabric, the cover, hand, and other aesthetics of finished fabrics made from texturized filament yarns resemble the properties of a fabric constructed from spun yarn.

Texturing is the process of crimping, imparting random loops, or otherwise modifying continuous filament yarn to increase cover, resilience, abrasion resistance, warmth, insulation, and moisture absorption, or to provide a different surface texture. Texturing methods fall roughly into the following six groups.

1. **Air-jet method:** In this method, yarn is led through the turbulent region of an air jet at a rate faster than it is drawn off on the far side of the jet. In the jet, the yarn structure is opened, loops are formed, and the structure is closed again. Some loops are locked inside the yarn and others are locked on its surface.
2. **Edge-crimping method:** In this method, thermoplastic yarns in a heated and stretched condition are drawn over a crimping edge and cooled. Edge-crimping machines are used to make Agilon yarns.
3. **False-twist method:** This continuous method for producing textured yarns utilizes simultaneous twisting, heat-setting, and untwisting. The yarn is taken from the supply package and fed at controlled tension through the heating unit, through a false-twist spindle or over a friction surface that is typically a stack of rotating discs called an aggregate, through a set of take-up rolls, and on to a take-up package. The twist is set into the yarn by the action of the heater tube and subsequently is removed above the spindle or aggregate, resulting in a group of filaments with the potential to form helical springs.
4. **Gear-crimping method:** In this method, yarn is fed through the meshing teeth of two gears. The yarn takes on the shape of the gear teeth.
5. **Knit-de-knit method:** In this method, the yarn is knit into a two-inch-diameter hose-leg, heat-set in an autoclave, and then unraveled and wound on to a final package. This texturing method produces a crinkle yarn.
6. **Stuffer box method:** The crimping unit consists of two feed rolls and a brass tube stuffer box. By compressing the yarn into the heated stuffer box, the individual filaments are caused to fold or bend at a sharp angle, while being simultaneously set by a heating device.

The physical, chemical, and mechanical properties of yarns are characterized in methods similar to those used for fibers. To describe the fineness of the yarn, textile engineers often use the unit tex, which is the weight in grams of a kilometer of yarn, or decitex, which is a finer measurement corresponding to the weight in grams of 10 km of yarn. Many other units have been used over time by different industries.

Yarn quantities are usually measured by weight in ounces or grams. In the United States, Canada, and Europe, balls of yarn for handcrafts are sold by weight. Common sizes include 25, 50, and 100 g skeins. Some companies primarily measure in ounces, with common sizes being three-ounce, four-ounce, six-ounce, and eight-ounce skeins. These measurements are taken at a standard temperature and humidity, because yarn can absorb moisture from the air. The actual length of the yarn contained in a ball or skein can vary due to the inherent heaviness of the fiber and the thickness of the strand. For instance, a 50 g skein of cotton yarn may contain several hundred meters, while a 50 g skein of bulky wool may contain only 60 m.

4.3 Fabric manufacturing techniques

In the textile industry the traditional methods of manufacturing fabrics from yarns are weaving, knitting, and braiding. More unconventional methods of fabric making include bonding fibers by mechanical, thermal, chemical, or solvent means.

4.3.1 Weaving

Weaving is described as interlacing, usually at right angles, two sets of threads to form cloth, rugs, or other types of woven textiles. In this process, two distinct sets of yarns called the warp and the filling, or weft, are interlaced with each other to form a fabric. The lengthwise yarns which run from the back to the front of the loom are called the warp and the crosswise yarns are the filling or weft. A loom is a device for holding the warp threads in place while the filling threads are woven through them.

Woven fabrics are used in the medical field for many applications. For example, blood filtration utilizes woven fabrics to allow blood cells to pass through while retaining larger particles such as blood clots, tissue fragments, and other debris. These fabric constructions are characterized by a large open area, high permeability, minimal blinding, and a straight flow path. Open-weave constructions are utilized down to a pore size of approximately 30 μm. The weave construction is changed if smaller openings are required. A closed-weave construction forms small openings while maintaining the fiber strength and integrity. Such fabrics are often chosen as an alternative to membranes and nonwovens, since the precision woven structures offer more consistent filter characteristics. These woven textile materials are non-shedding, pliable, and can be tailored to have wicking and absorption properties.

4.3.2 Knitting

Knitting is the most prevalent method of fabric construction after weaving. In the knitting process, the yarn follows a meandering path forming symmetric loops or stitches. When the interlocking loops run lengthwise, each row is called a wale. A wale can be compared with the warp in weaving. When the loops run across the fabric, each row is called a course. A course corresponds to the filling, or weft. There are two major varieties of knitting: weft and warp. In weft knitting, one continuous yarn forms courses across the fabric. In warp knitting, a series of yarns form wales in the lengthwise direction of the fabric. In general, knitted products are extensible, absorbent, and elastic, which are qualities required in many wound-care and implantable medical textile products.

4.3.3 Braiding

Braiding is one of the major fabrication methods, and produces a rope-like material by interweaving three or more strands, strips, or lengths in a diagonally overlapping pattern. It can be classified into two- and three-dimensional braiding. The two-dimensional braid structure can be circular or flat, and is formed by crossing a number of yarns diagonally so that each yarn passes alternately over and under one or more of the others. Two-dimensional braids are produced using circular braiding or rotary machines. Three-dimensional braiding is relatively new and was developed mainly for composite structures.

Because of their requirement for strength and flexibility, surgical sutures are often made with braided structures.

4.3.4 Nonwoven fabrics

Nonwoven fabrics are made by bonding or interlocking fibers or filaments by mechanical, thermal, chemical, or solvent means. When nonwoven fabrics are made from staple fibers, their production involves the formation of a uniform web by a wet-laid process or carding, followed by bonding the nonwovens either thermally or by other means such as needle punching, hydroentangling, etc. Spun-laid nonwovens are made in one continuous process where fibers are spun and then directly dispersed into a web by deflectors or air streams. Melt-blown nonwoven is a one-step process in which high-velocity air blows a molten thermoplastic resin from an extruder die tip on to a conveyor or take-up screen to form a fine fibrous and self-bonded web. Figure 4.2 shows a schematic illustration of the melt-blown process.

Because the fibers are loosely connected, nonwoven fabrics are often more porous than other forms of textiles. In order to gain integrity, the fibers in nonwoven fabric are bonded by using one of the following methods.

1. **Bonding with binder fibers:** Specially engineered low-melting-point fibers are blended with other fibers in a web, so that a uniformly bonded structure can be generated at low temperature by fusion of the binder fiber with adjacent fibers.
2. **Spot bonding:** This is a process of binding thermoplastic fibers into a nonwoven fabric by applying heat and pressure so that a discrete pattern of fiber bonds is formed.

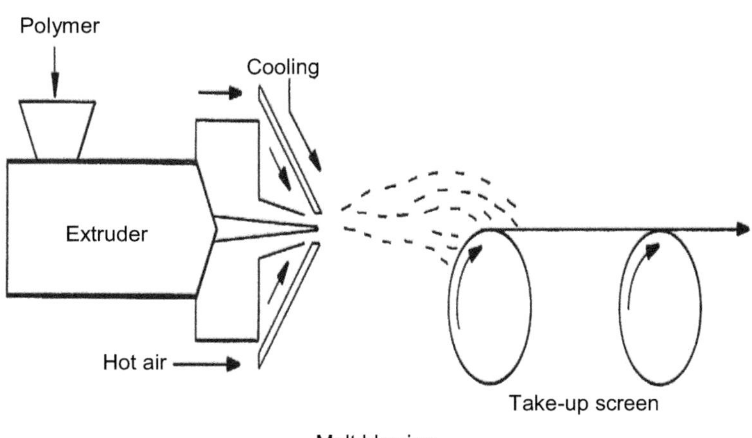

Figure 4.2 Schematic illustration of the melt-blown process.

3. **Print bonding:** Adhesive material is printed on the nonwoven fabric in a discrete pattern to achieve bonding of the fibers.
4. **Saturation bonding:** By soaking the web with an adhesive, the fibers in a nonwoven fabric can be bonded together.
5. **Spray bonding:** This is a process of binding fibers into a nonwoven fabric using spray application of a fabric binder.
6. **Stitch bonding:** In this method, the fibers are connected by stitches sewn or knitted through the web.

4.4 Treatment of textile fabrics

Textile fabrics receive a number of chemical and physical treatments before they are made into a final product. From formaldehyde finishes to improve crease resistance to flame-retardant treatment and dyeing of many types of fabrics, the possibilities for the application of dyeing and finishing processes to change the characteristics of textile materials are almost endless. Recent advances in nanomaterials have led to the development of treatments based on metallic nanoparticles for making textiles more resistant to water, stains, wrinkles, and pathogens such as bacteria and fungi.

For medical textile materials, many of the dyeing and finishing agents may have detrimental effects on the end users. For example, a number of disperse, acid, and reactive dyes have been shown to be allergenic to sensitive individuals. Some residue finishing agents may cause allergic responses when in direct contact with the human body. It is therefore important to select the appropriate processes and raw materials for the treatment of textile materials, especially those for medical applications.

4.4.1 Bleaching

Bleach refers to a number of chemicals that remove color, whiten, or disinfect, often via oxidation. Although the bleaching process has been known for thousands of years, the chemicals currently used for bleaching resulted from the work of several eighteenth-century scientists, including Swedish chemist Carl Wilhelm Scheele, who discovered chlorine, and French scientists Claude Berthollet, who recognized that chlorine could be used to bleach fabrics and first made sodium hypochlorite, and Antoine Germain Labarraque, who discovered the disinfecting ability of hypochlorites. Scottish chemist and industrialist Charles Tennant first produced a solution of calcium hypochlorite, and then solid calcium hypochlorite (bleaching powder). Louis Jacques Thénard first produced hydrogen peroxide in 1818 by reacting barium peroxide with nitric acid. Hydrogen peroxide was first used for bleaching in 1882, but did not become commercially important until after 1930.

Chlorine is the basis for the most commonly used bleaches, for example the solution of sodium hypochlorite, which is so ubiquitous that many people just call it "bleach," and calcium hypochlorite. Oxidizing bleaching agents that do not contain chlorine are often based on peroxides, such as hydrogen peroxide, sodium percarbonate, and sodium perborate.

While most bleaches are oxidizing agents, a lesser number are reducing agents, such as sodium dithionite. Chemically, color in most dyes and pigments is produced by molecules which contain chromophores. To remove color, chemical bleaches work in one of two ways.

- An oxidizing bleach works by breaking the chemical bonds that make up the chromophore, which changes the molecule into a different substance that either does not contain a chromophore or contains a chromophore that does not absorb visible light.
- A reducing bleach works by converting double bonds in the chromophore into single bonds. This eliminates the ability of the chromophore to absorb visible light.

Sunlight can act as a bleach, since high-energy photons of light, often in the violet or ultraviolet range, can disrupt the bonds in the chromophore, rendering the resulting substance colorless. Extended exposure often leads to massive discoloration, usually reducing the colors to white and faded-blue spectrums.

In addition to whitening fabric, the broad-spectrum antimicrobial effectiveness of bleach, particularly sodium hypochlorite, is due to the nature of its chemical reactivity with microbes. Rather than acting in an inhibitory or toxic fashion in the manner of antibiotics, bleach quickly reacts with microbial cells to denature and destroy many pathogens irreversibly. Bleach, particularly sodium hypochlorite, has been shown to react with a microbe's heat-shock proteins, causing the bacteria to form into clumps that will eventually die off. In some cases, the base acidity of a bleach compromises a bacterium's lipid membrane—a reaction similar to popping a balloon. The range of microorganisms effectively killed by bleach is extensive, making it an extremely versatile disinfectant.

4.4.2 Dyeing

Textile fabrics are available in almost every color and pattern. Colored designs in textiles can be created by weaving together fibers of different colors (such as Scottish tartan or Uzbek ikat), adding colored stitches to finished fabric (such as embroidery), creating patterns by resist-dyeing methods, tying off areas of cloth and dyeing the rest (tie-dyeing), drawing wax designs on cloth and dyeing in between them (batik), or using various printing processes on finished fabric.

Dyeing is the process of adding color to textile products such as fibers, yarns, and fabrics. Dyeing is normally done in a special solution containing dyes and other chemicals so that the dye molecules can penetrate into the fiber structure. Temperature and time controlling are two key factors in dyeing.

There are mainly two classes of dye: natural and synthetic. For most of the thousands of years in which dyeing has been used by humans to decorate clothing or fabrics, the primary source of dye has been nature, with the dyes being extracted from animals or plants. In the last 150 years humans have produced artificial dyes to achieve a broader range of colors and render the dyes more stable to resist washing and general use.

Different classes of dyes are used for different types of fiber and at different stages of the textile production process, from loose fibers through yarn and cloth to completed garments. For example, acrylic fibers are dyed with basic dyes, nylon and protein fibers such as wool and silk are dyed with acid dyes, and polyester yarn is dyed with disperse dyes. Cotton is dyed with a range of dye types, including vat dyes, reactive dyes, and direct dyes.

Dyes are applied to textile goods by dyeing from dye solutions and printing from dye pastes.

4.4.3 Finishing

To impart the required functional properties to fibers, yarns, and fabrics, it is customary to subject textile materials to different types of physical and chemical treatments. For example, wash-and-wear finish for a cotton fabric is necessary to make it crease- or wrinkle-free. In a similar way, mercerizing, singeing, flame-retardant, water-repellent, waterproof, antistatic, and peach finishing achieve various fabric properties desired by consumers.

Finishing refers to any process performed on textile materials to improve the look, performance, or hand of the finished textile or clothing. Some finishing techniques, such as fulling, have been in use with handweaving for centuries; others, such as mercerization, are byproducts of the Industrial Revolution.

In general, the properties of plastic-based synthetic fibers—most important among them being polyamide, polyester, and polyacrylonitrile—are essentially different from those of natural cellulosic and wool fibers, hence the sequence of finishing operations is likely to be different. For example, while cellulosic fabrics require a resin finishing treatment to impart easy-care properties, synthetic fibers already exhibit these easy-care criteria and require only a heat-setting operation.

In the textile industry there are three types of standard finishes.

- Quality-oriented finishing, such as calendering, decatizing, desizing for woven fabrics, pressing, scouring with detergents, alkaline solutions, or enzymes to remove foreign matter, shrinking, sanforization, shearing, or singeing to smooth the fabric by removing the fine protruding fibers on its surface. Flame singeing is the standard process where the wet fabric is passed through an array of gas burners at a suitable distance to burn the pills off its surface.
- Design-oriented finishing, such as bleaching of woven fabrics to remove any prior color in order to obtain a uniform color during the dying process. The color and pattern of textile materials in a particular design can be achieved by dyeing, printing, and other finishing techniques.
- Handle-oriented finishing, such as fulling to add weight and density, hydrophobic finishing to produce a fabric that repels stains or water, and weighting silk with metallic salts or polymer to add weight and improve handle.

Some of the commonly used finishing processes are described below.

1. Biopolishing removes the protruding fibers of a fabric through the action of an enzyme. Enzymes, such as cellulase for cotton, selectively remove protruding fibers. These enzymes may be deactivated by an increase in temperature.
2. Mercerization makes woven cotton fabric stronger, more lustrous, have better dye affinity, and be less abrasive.
3. Raising lifts the surface fibers to improve softness and warmth, as in flannelette.

4. Peach finish subjects the fabric to emery wheels, making the surface velvet-like.
5. Fulling is a method of thickening woolen material to make it more water resistant.
6. Decatizing brings dimensional stability to woolen fabrics.
7. Calendering makes one or both surfaces of the fabric smooth and shiny by passing the fabric through hot, fast-moving, stainless steel cylinders.
8. Sanforizing or preshrinking prevents a fabric and the garments produced from it from shrinking after production. This is a mechanical finish, acquired by feeding the fabric between a roller and a rubber blanket in such a way that the rubber blanket compresses the weft threads and imparts compressive shrinkage.
9. Crease-resistant, "wash-and-wear," or "wrinkle-free" finishes are achieved by the addition of a chemical resin finish that makes a natural fiber take on a quality similar to that of synthetic fibers.
10. An antimicrobial finish causes a fabric to inhibit the growth of microbes. The humid and warm environment found in textile fibers encourages the growth of microbes. Infestation by microbes can cause cross-infection by pathogens and the development of odor where the fabric is worn next to skin. An antimicrobial finish is applied to textile materials to protect the skin of the wearer and the textile substrate itself.

4.4.4 Printing

Printing is a process for producing a pattern on yarns, warp, fabric, or carpet. The color or other treating material, usually in the form of a paste, is deposited on to the fabric, which is then usually treated with steam, heat, or chemicals for fixation. There are many types of printing methods, some of which are described below.

Block printing: The printing of fabric by hand, using carved wooden or linoleum blocks, as distinguished from printing by screens or roller.

Blotch printing: A process wherein the background color of a design is printed rather than dyed.

Burn-out printing: A method of printing to obtain a raised design on a sheer ground. The design is applied with a special chemical on to a fabric woven of pairs of threads of different fibers. One of the fibers is then destroyed locally by chemical action. Burn-out printing is often used on velvet. The product of this operation is known as a burnt-out print.

Direct printing: A process wherein the colors for the desired designs are applied directly to white or dyed cloth, as distinguished from discharge printing and resist printing.

Discharge printing: In "white" discharge printing the fabric is piece dyed, then printed with a paste containing a chemical that reduces the dye and hence removes the color where white designs are desired. In "colored" discharge printing a color is added to the discharge paste in order to replace the discharged color with another shade.

Duplex printing: A method of printing a pattern on the face and the back of a fabric with equal clarity.

Heat-transfer printing: A method of printing fabric of polyester or other thermoplastic fibers with disperse dyes. The design is transferred from preprinted paper on to the fabric by contact heat, which causes the dye to sublime. Having no affinity for paper, the dyes are taken up by the fabric. The method is capable of producing well-defined, clear prints.

Ink-jet printing: Noncontact printing that uses electrostatic acceleration and deflection of ink particles released by small nozzles to form the pattern.

Photographic printing: A method of printing from photoengraved rollers. The resultant design looks like a photograph. The designs may also be photographed on a silk screen, which is used in screen printing.

Pigment printing: Printing by the use of pigments instead of dyes. The pigments do not penetrate the fiber but are affixed to the surface of the fabric by means of synthetic resins which are cured after application to make them insoluble. The pigments are insoluble, and application is in the form of water-in-oil or oil-in-water emulsions of pigment pastes and resins.

Resist printing: A printing method in which the design can be produced by applying a resist agent in the desired design and then dyeing the fabric, in which case the design remains white although the rest of the fabric is dyed; or by including a resist agent and a dye in a paste which is applied for the design, in which case the color of the design is not affected by subsequent dyeing of the fabric background.

Roller printing: The application of designs to fabric using a machine containing a series of engraved metal rollers positioned around a large padded cylinder. Print paste is fed to the rollers and a doctor blade scrapes the paste from the unengraved portion of the roller. Each roller supplies one color to the finished design, and as the fabric passes between the roller and the padded cylinder, each color in the design is applied. Most machines are equipped with 8 rollers, although some have 16 rollers.

Rotary-screen printing: A combination of roller and screen printing in which a perforated cylindrical screen is used to apply color. Color is forced from the interior of the screen on to the cloth.

Screen printing: A method of printing similar to using a stencil. The areas of the screen through which the coloring matter is not to pass are filled with a waterproof material. The printing paste which contains the dye is then forced through the untreated portions of the screen on to the fabric below.

Warp printing: The printing of a design on the sheet of warp yarns before weaving.

4.5 Packaging and sterilization

Because medical textile materials are used in a medical environment, quite often in direct contact with the human body, it is important to ensure that these materials are sterile prior to application, that is, they are free of bacterial contamination. Sterility of the finished product is achieved in two stages: first, medical textile materials are packed in appropriate packaging; and second, they are sterilized with an appropriate method.

Because of the wide variety of products in the medical textiles industry, packaging requirements differ from product to product, and a wide range of materials can be used to pack medical textile products. For the production of wound dressings, individual products are first packed in pouches before being packed into cartons and boxes. The pouches are often made from a combination of paper, polymer film, and aluminum foil to achieve the required level of protection against bacterial contamination, light, moisture, etc. In addition, the packaging processes are carried out in a clean environment to ensure the bioburden of the product meets requirements prior to sterilization.

Sterilization is a term referring to any process that eliminates or kills all forms of microbial life, including transmissible agents such as fungi, bacteria, viruses, spore, etc., which are present on a surface or contained in a fluid or a compound such as a medical textile material. Sterilization can be achieved by applying heat, chemicals, irradiation, high pressure, or filtration. In the medical textile industry, the main methods used to sterilize products are heat, ethylene oxide, gamma irradiation, and electron radiation.

4.5.1 Steam sterilization

Steam is widely used in heat sterilization, where an autoclave is filled with steam heated to 121–134 °C (250–273 °F). To achieve sterility, a holding time of at least 15 min at 121°C (250 °F) at 100 kPa (15 psi) or 3 min at 134 °C (273 °F) at 100 kPa (15 psi) is required. Additional sterilizing time is usually required for liquids and instruments packed in layers of cloth, as they may take longer to reach the required temperature. Following sterilization, liquids in a pressurized autoclave must be cooled slowly to avoid boiling over when the pressure is released.

4.5.2 Ethylene oxide

The chemical formula of ethylene oxide (EO or EtO) is C_2H_4O. It is commonly used to sterilize objects such as plastics, optics, and electrics, which are sensitive to temperatures greater than 60 °C and/or radiation. EtO treatment is generally carried out between 30 °C and 60 °C, with relative humidity above 30% and a gas concentration between 200 and 800 mg/l, and typically lasts for at least 3 h. EtO penetrates well by moving through paper, cloth, and plastic films and is highly effective against all known viruses, bacteria, and fungi, including bacterial spores. It is also compatible with most materials, even when repeatedly applied. However, EtO is highly flammable, toxic, and carcinogenic, with a potential to cause adverse reproductive effects.

A typical sterilization process with EtO consists of a preconditioning phase, the actual sterilization run, and a period of poststerilization aeration to remove toxic residues, such as EtO, ethylene glycol (formed out of EtO and ambient humidity), and ethylene chlorohydrine (formed out of EtO and materials containing chlorine, such as PVC). In the medical device industry EtO is the most common sterilization method, used for over 70% of total sterilizations and 50% of all disposable medical devices.

4.5.3 Radiation sterilization

Radiation sterilization involves the exposure of medical textile products to radiations such as electron beams, X-rays, gamma rays, or subatomic particles. Gamma rays are very penetrating and are commonly used for the sterilization of disposable medical devices, such as syringes and needles. The gamma radiation is emitted from a radioisotope, usually cobalt-60 or caesium-137. Caesium-137 is used in small hospital units to treat blood before transfusion to prevent graft-versus-host disease. The use of a radioisotope requires shielding for the safety of the operators while in use and in storage, as the radioisotope continuously emits gamma rays.

Electron beams are also commonly used for sterilization, and provide a much higher dosing rate than gamma or X-rays. Due to the higher dose rate, less exposure time is needed and thereby any potential degradation of polymers is reduced. A limitation is that electron beams are less penetrating than either gamma or X-rays. On the contrary, X-rays are highly penetrating and can be used to irradiate large packages and pallet loads of medical devices. Their penetration is sufficient to treat multiple pallet loads of low-density packages with very good dose uniformity ratios. In addition, X-ray sterilization is an electricity-based process that does not require chemical or radioactive materials. High-energy and high-power X-rays are generated by an X-ray machine that can be turned off when not in use, and therefore the sterilization process does not require any shielding when equipment is in storage.

4.6 Summary

As a novel type of textile material, medical textiles are made in ways similar to traditional textile materials, and the many textile technologies such as weaving, knitting, braiding, nonwoven processing, bleaching, dyeing, finishing, and printing are all highly valuable in the manufacture of modern textile products for medical applications. For medical textiles these processes are carried out on a small scale, in a clean environment, and often in combination with downstream operations such as packaging and sterilization with a suitable method.

Sources for further information

[1] Atkinson C. False twist textured yarns: principles, processing and applications. Cambridge: Woodhead Publishing Ltd; 2012.

[2] Lawrence CA. Advances in yarn spinning technology. Cambridge: Woodhead Publishing Ltd; 2010.

[3] Lord PR. Handbook of yarn production: technology, science and economics. Cambridge: Woodhead Publishing Ltd; 2003.

[4] Chapman R. Applications of nonwovens in technical textiles. Cambridge: Woodhead Publishing Ltd; 2010.

[5] Ormerod A, Sondhelm WS. Weaving: technology and operations. Cambridge: Woodhead Publishing Ltd; 1995.

[6] Schindler WD, Hauser PJ. Chemical finishing of textiles. Cambridge: Woodhead Publishing Ltd; 2004.

[7] Rajendran S, editor. Advanced textiles for wound care. Cambridge: Woodhead Publishing Ltd; 2009.

[8] Qin Y. Review on advanced wound dressings. J Text Inst 2001;92:127–38.

Applications of advanced technologies in the development of functional medical textile materials

5

5.1 Introduction

Textile materials include fibers, filaments, yarns, fabrics with woven, knitted, and nonwoven structures, and many other forms of materials and articles made from natural and synthetic fibrous substances. In addition to their traditional applications in the clothing and garment industries, textile materials are finding more and more novel end uses, among which medical applications represent a fast-expanding industry. A number of factors contribute to the demand for these textile-based products, such as aging populations and the hazards of human activities, including transport accidents, chemical injuries, fire, cold, diseases, sports, etc. Such factors stimulate the rapid development of the medical textile market in using novel techniques and technologies to generate modern textile materials and their related products.

Textile products have many unique characteristics and excellent qualities, such as strength, extensibility, flexibility, air and moisture permeability, availability in three-dimensional structures, variety in fiber length, fineness, cross-sectional shape and geometry, and absorbency, etc. Because of these novel properties, textile materials are ideally suited for the manufacture of products in fields such as wound healing and controlled drug release, bandaging and pressure garments, implantable materials, and other forms of medical devices. In addition, advances in textile technologies mean that nowadays textile products are able to combine traditional textile characteristics with modern multifunctionality. High surface areas, absorbency properties, and large varieties in product forms contribute to the emergence of more and more smart products in the medical textile sector.

In addition to their own unique structures and product performances, textile materials can be combined with other types of materials to produce composites with enhanced functionality. In this respect, developments in related fields such as nanomaterials, biomaterials, microelectronics, biotechnology, etc., have made it possible to develop smart textile materials for medical and healthcare applications by utilizing functional biomaterials, photochromic materials, thermochromic materials, conductive polymers, and fibers, as well as phase-change materials, light-emitting polymers, optical fibers, shape-memory polymers, etc. A number of fabrication forms have also been developed—for example, the fabrication forms of shape-memory polymers include films, springs, clamps, foils, and wires which can be used as components of textile structures. In addition, developments in tissue engineering and nanotechnologies have had a great impact on advanced medical textile products in areas such

Medical Textile Materials. http://dx.doi.org/10.1016/B978-0-08-100618-4.00005-4

as sutures, joint replacements, artificial ligaments, tendons, wound dressings, blood vessels, heart valves, artificial skin, scaffolds enabling tissue regeneration, protective structures, biomonitoring systems, etc.

In response to ever-changing societal needs, there is a sustained demand for medical textile professionals to find better solutions to meet clinical requirements from patients and carers. In this respect, advances in medical textile materials will originate from the new developments in polymer, fiber, textile, and more importantly composite technologies. These developments cover a wide range of industries, and some of these closely related to the medical textile industry are discussed below.

5.2 Polymer engineering for advanced medical and healthcare products

As the basic component of medical textile materials, the structures and properties of the constituent polymers have a significant effect on the biodegradability, biocompatibility, absorbency, antimicrobial property, and other functional performances of the final medical textile products. Functional modifications of polymers have far-reaching effects on the fibers, yarns, fabrics, and textile materials that are processed in a series of downstream operations. In order to generate the desired product performance characteristics for their diverse applications such as hygiene, protection, therapeutic, non-implantable or implantable materials, extracorporeal devices, etc., the chemical and physical structures of the relevant polymers should be engineered to suit their required specifications.

For natural polymers, the variations in raw material supplies and extraction processes mean that their performance characteristics often vary over a wide range. For example, alginate extracted from different types of seaweed differs in monomer composition, molecular weight, and impurity contents. Chemical modification of alginate is often used as a tool to enhance existing properties, such as the improvement of ionic gel strength by additional covalent cross-linking, improvement in biodegradability, etc. In addition, the alteration of its natural chemical structure can generate completely new properties otherwise not existing in unmodified alginates—for example, anticoagulant properties, chemical and biochemical anchors to interact with cell surfaces, and temperature-dependent characteristics such as lower critical solution temperature. In short, alginate derivatization is the most convenient way to achieve both inherent property enhancement and new property introduction.

As shown in Figure 5.1, when treated with $NaIO_4$ the hydroxyl groups in alginate are oxidized to generate a water-soluble polymer with aldehyde groups. This alteration in the chemical structure of alginate has a significant effect on its applications. Since aldehyde groups can easily react with amine groups, oxidized alginate can be used to treat chitosan fibers to enhance absorption properties by attaching water-holding sodium carboxymethyl groups to the fiber surface. In addition, when reacted with propylene oxide the resultant propylene glycol alginate can be mixed with chitosan in an

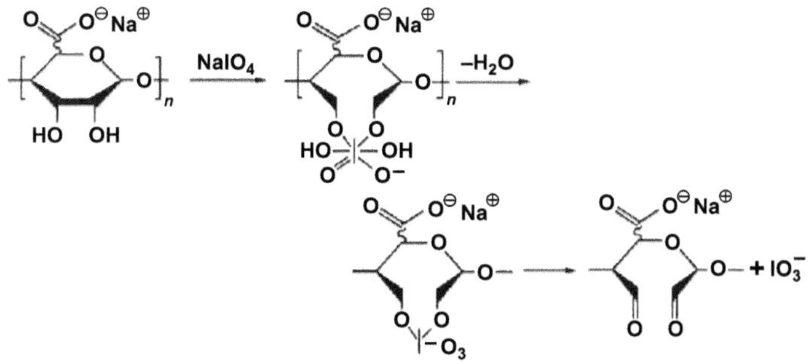

Figure 5.1 The aldehyde structure of oxidized sodium alginate.

acidic solvent to obtain a homogeneous spinning solution. The resultant fiber contains both alginate and chitosan uniformly distributed in its structure.

Similar to the chemical modifications applied to alginate, chitosan can be treated with chloro-acetic acid, ethylene oxide, and propylene oxide to obtain sodium carboxymethyl chitosan, hydroxyethyl chitosan, and hydroxypropyl chitosan, respectively. These derivatives are highly absorbent and can be used for the production of superabsorbent fibers for medical applications. In addition to enhanced absorption properties, it is known that carboxymethyl chitosan possesses unique wound-healing properties whereby the application of foams, hydrogels, or nonwoven dressings can effectively promote the healing of chronic ulcerative wounds.

For synthetic polymers, the structures and properties can be easily changed by using different monomers and changing the composition of monomers used for polymerization. For example, polyglycolic acid or polyglycolide (PGA) is known as a fiber-forming polymer which can be polymerized directly from glycolic acid, while polylactic acid or polylactide (PLA) is polymerized from lactic acid. Because of the extra methyl group in the PLA repeating unit, PLA is more hydrophobic than PGA. To improve the properties of PLA and PGA, copolymers of PLA and PGA (PLGA) can be prepared by using lactic acid and glycolic acid as the monomers during polymerization. Materials made from PLGA have a shorter absorption time when compared with PLA. In addition, the degradation time can be tailored by altering the ratio of the monomers used during the synthesis of PLGA for the manufacture of healthcare products such as skin grafts, implant materials, prosthetic devices, sutures, microparticles, and nanoparticles.

In general, both natural and synthetic polymers can be chemically modified to generate improved chemical, physical, and biological activities. The bioactivity of polymeric materials is particularly important in implantable devices and scaffolds for tissue regeneration. In order to provide polymers with bioactive properties, the chemical structures of the polymer can be modified by changing the monomer composition, altering the main chain structure by grafting of side chains, and chemically converting some functional groups on the polymer chains. In this respect, significant advances have been made in the syntheses and applications of polymers to prevent microbial attack and degradation by incorporating antimicrobial elements into the polymer structure. This can be achieved by using monomers containing biocide moieties followed by the polymerization of these monomers alone or copolymerizing with other monomers, or by grafting antimicrobial agents on to natural or synthetic polymers.

5.3 Fiber engineering for advanced medical and healthcare products

As fibers are the basic structural component of textiles, their chemical, physical, mechanical, and biological properties are closely related to the structural, physical, and mechanical characteristics of medical textile materials, such as stable and spatial structure, purity and nontoxicity, sterility, ability to manage exudates and fluid without causing irritation or maceration, cushioning effect, comfort, pain relief, provision of thermal insulation, breathability, mechanical protection, moisture and liquid absorption, low adherence, nonsensitizing, nonallergic, not contaminating a wound with loose fibers or other particles, providing an effective barrier against microorganisms, dirt, liquid, and other foreign bodies, etc. To meet performance requirements, many engineering methods have been developed to alter the structural, physical, and mechanical characteristics of textile fibers. Some of the commonly used techniques are described below.

5.3.1 Microfibers

Microfibers are fibers that are less than one denier per filament—100 times finer than a human hair. Textiles produced with microfiber yarns are softer and possess better drape than conventional products. Microfibers can be made from polyester, polyamide, polypropylene, and many other fiber-forming polymers. These fibers can be made into fabrics with dimensionally stable, lightweight, waterproof, and easy-care characteristics.

5.3.2 Nanotechnology and engineering

The "nano" in nanotechnology refers to materials with particle sizes of less than one-billionth of a meter. In the textile industry, nanotechnology can be used to provide performance characteristics such as antimicrobial and antistatic properties, UV resistance, stain resistance, wrinkle resistance, etc. In recent years advanced technologies have been developed to enable the production of nanofibers, nanotubes, and nanorods from synthetic and natural polymers. In addition, traditional fibers can be modified by the introduction of functional nanoparticles such as superparamagnetic particles, antimicrobial agents, enzymes, etc. Such "value-added" nanofibers can be used effectively in protective clothing, drug delivery, tissue engineering, etc.

As an example, to give functional fibers bioactivities, Cupron Inc. utilized the bioactivities of copper to develop a new type of copper-containing antimicrobial fiber. During the melt-spinning process, copper oxide nanoparticles were mixed into the melt and dispersed in the fiber. This type of copper-containing fiber uses the ability of copper to suppress the growth of bacteria, viruses, and fungi and has good antibacterial and antiviral properties. Because copper can stimulate collagen formation in tissue, copper-containing fibers can promote skin regeneration to make healthy skin smoother and ensure fast healing of wounded skin. They are highly valuable in the development of functional wound dressings, antimicrobial textile products, and other related fields.

In another example, German company Zimmer blended seaweed nanoparticles with the spinning solution during the solvent spun process for cellulose fibers, and dispersed the seaweed particles in the resultant Seacell fibers, which were further treated with aqueous silver nitrate to incorporate silver ions into the fibers through its absorption by the seaweed particles. The silver-containing Seacell Active fibers have good antimicrobial properties due to the antimicrobial effect of silver, and have been effective against *Candida albicans*, *Staphylococcus aureus*, and *Escherichia coli* that are responsible for itching, dermatitis, and skin infection respectively. Because the Seacell and Seacell Active fibers combine the excellent physical properties of Lyocell fibers and the bioactivities of seaweed biomaterials as well as silver ions, they can be used in the production of functional medical textile materials.

5.3.3 Electrospinning

In the electrospinning process, a strong electrical field is applied to a droplet formed by a polymer solution or polymer melt at the tip of a die acting as one of the electrodes. The counter electrode is located over a distance. The charging of the fluid leads eventually to the ejection of a jet from the tip of the cone. The charged jet is accelerated toward the counter electrodes and thins rapidly during this period due to elongation and evaporation of the solvent, until solid fibers are deposited on to the substrate located on top of the counter electrode in a random fashion.

The electrospinning process is characterized by a rapid evaporation of the solvent in solution spinning and a rapid temperature decrease in spinning from the melt. Although structure formation happens at a millisecond scale, the degree of crystallinity and the perfection of the crystals which grow within the fibers are similar to those observed for thicker fibers as obtained, for instance, from melt extrusion. Another feature is the strong deformation taking place during the whipping mode, which tends to give rise to orientational processes within the fibers. High degrees of crystal orientation have been observed for various electrospun polymers, including polyethylene oxide and polyamide.

Coelectrospinning is a new form of the electrospinning process which aims at producing core-shell nanofibers. In this method, two liquids of different chemical compositions emerge from a core and the surrounding concentric annular nozzles, respectively. The droplet formed at the edge of such nozzles has one type of liquid encircling another, which forms a shell encircling a core when the droplet is pulled by the electric field, stretched, and solidified as the solvents evaporate during the spinning process. Since the coelectrospinning process is fast enough to prevent any mixing of the core and shell polymers, it is possible for the shell polymer to carry core materials which will not form fibers via electrospinning by themselves, such as dispersions of drugs. In this sense, the shell polymer serves as a template for the core material, leading to compartment-type structures. Core-shell fibers of this type are useful in many applications such as in the field of slow-release drug delivery.

A broad range of polymers can be electrospun into nanofibers from either solution or melt. One advantage of electrospinning is that water can be used as a solvent, for example during the preparation of fibers from polyethylene oxide and polyvinyl alcohol. Organic solvents are used during the electrospinning of fibers from PLA, polystyrene, polyamides, polyimides, and many other polymers. These polymers can be mixed with drugs or other biologically active molecules in the solution or melt prior to electrospinning, so that composite nanofibers can be made to carry the bioactive ingredients.

5.3.4 Gelling fibers

Gelling fibers are fibers that can form a fibrous hydrogel when in contact with an aqueous media. They can absorb a large amount of water into the fiber structure while maintaining this fibrous structure, which is important to obtain an integral piece of hydrogel during applications. Wound dressings made from this type of fibers have very high absorption capacities and can form a layer of fibrous gel on the wound surface, which provides a moist healing environment that can effectively speed up the healing of wounds.

There are many ways to engineer gelling fibers from various types of polymers. For example, carboxymethylation treatment can be applied to cellulosic fibers such as cotton, viscose rayon, and chitosan fibers to make carboxymethyl cellulose and carboxymethyl chitosan fibers respectively. Similarly, alginate fibers can be made highly absorbent by introducing sodium carboxymethyl cellulose into the fiber structure. When hydrophilic groups are introduced into their structure, gelling fibers can hold a large amount of water inside the polymeric matrix within the fibers, thereby enhancing their ability to sustain a moist but not the wet environment on the wound. Figure 5.2 shows the gelling process of a carboxymethylated cotton fiber.

5.3.5 Hollow fibers

Hollow fibers are widely used for filtration, utilizing the semipermeable nature of their capillary walls. In the medical industry, hollow fiber bioreactors are often made from cellulose and synthetic polymers. Cellulose acetate and cuprammonium rayon are the widely used cellulose-based hollow fibers, while synthetic hollow fibers are often made from polysulfone, polyamide, and polyacrylonitrile. Modifications can be made to these materials to improve their functions by using polymers based on phospholipid, a substance found in the human cell membrane. 2-methacryloyloxyethyl phosphorylcholine (MPC) is a methacrylate monomer with a phospholipid polar group. When MPC-based copolymers are used as additives for polysulfone, protein adsorption and platelet adhesion can be effectively reduced, thereby improving blood compatibility. Cellulose acetate hollow fiber membranes can also be modified with MPC-based copolymers by means of blending or surface coating to obtain improved permeability,

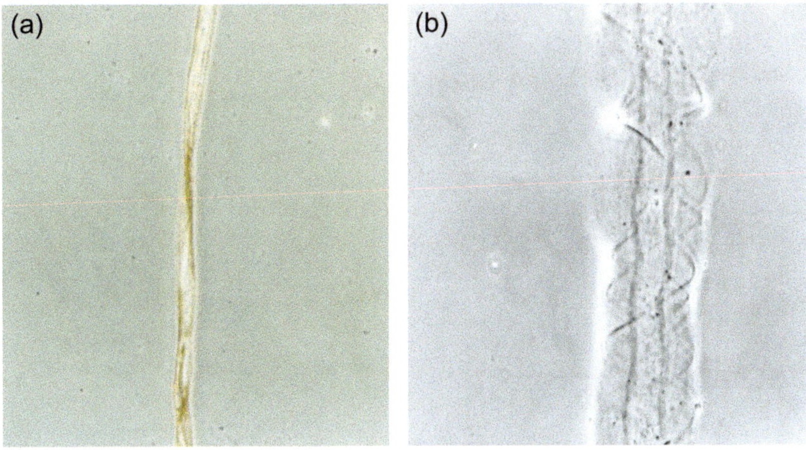

Figure 5.2 The dry (a) and wet (b) structures of carboxymethylated cotton fiber.
Source: Smart Textiles for Medicine and Healthcare (Chapter 2: Smart woundcare materials). Woodhead Publishing, 2007.

hemocompatibility, and cytocompatibility, which are important in hemopurification and liver-assist bioreactors. When hollow-fiber-based medical filters are used in direct contact with living cells or tissues, a specialized treatment is often required to improve the biocompatibility between the cells and the fiber surface. Various types of carbohydrate-immobilized phosphorylcholine polymers have been used as interfaces with living cells.

5.4 Textile engineering for advanced medical and healthcare products

Textile structures have a unique shape, porosity, and volume that are useful in many medical applications. Because of the wide range of technologies available, the structural characteristics of medical textile materials can also be engineered to offer desired performances. For example, an embroidery technique can be used to develop textile scaffold structures for tissue engineering, while three-dimensional embroidered textile architectures have a novel porous structure that is useful in wound-care applications. Other smart materials can be incorporated into the textile structures to aid recording and transmitting of biophysiological signals during diagnostics.

Some of the many textile engineering methods applicable for medical textile materials are described below.

5.4.1 Spacer fabrics

Spacer fabrics contain a combination of textile sheets and distance fibers, with a porous structure that can offer high absorption capacities for medical textile materials. Since the distance fibers are often monofilament polyester fibers with a marked stiffness, spacer fabrics can resist high pressure and have a directed transport of fluid and heat. The pressure-resistance property can also be adjusted by adjusting the type of fibers, the angle of the fibers in the textile structure, and the stitch density.

The mechanical and microclimatic properties of spacer fabrics are useful in compression bandages and bedding materials for decubitus prophylaxis. Spacer-fabric-based compression bandages are comfortable to wear, and when used on patients skin sweating is prevented due to the microclimatic quality of the spacer fabrics.

5.4.2 Surface treatment

The surface properties of textile materials can be manipulated and enhanced by implementing surface finishing, coating, and other techniques such as nanotechnology. For example, antibacterial and disinfection finishing techniques have been developed for many types of textiles by treatment with nanosized silver, where the treated products show effective antimicrobial activities against bacteria, fungi, and other microorganisms. In the wound management industry, a silver-coated nylon fabric has been widely used for the treatment of infected wounds.

To improve the moisture interaction properties of medical textile products, plasma surface treatment can be used to modify the absorptive characteristics of the textile structure. The treated products have superior moisture-wicking ability which can assist the transfer of moisture in vapor or liquid form quickly from the body into the textile structure.

5.4.3 Coating and lamination

Film-forming polymers can be coated on a textile substrate to enhance its performance. For example, chitosan can be coated on to nonwoven fabrics to produce nonwoven wound dressings with enhanced properties. In this method, chitosan is first dissolved in an acidic solution before the nonwoven fabric is immersed, and a layer of chitosan is formed on the nonwoven fabric by using the immersion-precipitation phase-inversion method. The chitosan-coated nonwoven wound dressing can control water loss due to evaporation, promote fluid drainage, and inhibit the invasion of exogenous microorganisms.

In the medical industry, porous thermoplastic or thermoset membranes can be used to remove extremely small particles. They are used in applications requiring surface filtration where the particles are trapped on the surface of the media. Although the polymeric membranes are able to remove particles ranging in size from the smallest ions ($<0.001\ \mu m$) up to approximately one micron particles, they tend to be expensive and fragile. By coating the membrane on a textile substrate or lamination of a polymeric membrane with a textile fabric, it is possible to combine the physical strength of the textile materials with the filtration properties of membrane materials. In this sense, the porous structure of hydrophobic polytetrafluoroethylene membranes is often used to construct medical textile materials that can be used for filtration purposes.

5.4.4 Other engineering methods

In the medical textile industry, the specific requirement of textile materials can be met by adopting many specialized engineering techniques. For example, transmission is an important feature for medical textile products, and is closely related to a range of characteristics such as air and water permeability, thermal resistance, cool/warm feeling, etc. Light mesh knit used in sports garments can also improve heat transfer and maximum body comfort for surgical and dental gowns. Soft silicone mesh is a nonadherent, porous dressing with a wound contact layer consisting of a flexible polyamide net coated with silicone. Its porous nature allows fluid to pass to the secondary dressing, while the silicone layer reduces adherence between the dressing and the wound surface.

Many nonwoven structures used for absorbent wound dressings exhibit anisotropic fluid transmission characteristics in the vertical and lateral directions of the fabric structure. The orientation of fibers within these materials is the main factor affecting anisotropic fluid transmission, and can be manipulated to design desirable transport properties by using appropriate fabric porosity, fiber diameter, and orientation distribution.

5.5 Modern composite technologies for advanced medical and healthcare products

Composite materials are made from two or more constituent materials with significantly different physical or chemical properties that, when combined, produce a material with characteristics different from the individual components. In the medical textile industry, modern composite technologies are often used to enhance the performance and functionalities of fibers, yarns, and textile materials. For example, silver-impregnated fibers and textiles are now widely used in wound dressings for infected wounds and wounds at high risk of infection. The silver component can be added into a wound dressing in a number of ways, such as by blending silver-coated nylon fibers with calcium alginate fibers before the mixture is carded and made into a needle-punched nonwoven fabric. In another example, a strain sensor based on conducting polymers or carbon-filled rubbers is incorporated into the textile structure to create wearable devices able to follow the posture and movements of a subject wearing the system, for example, when under rehabilitation. Garments that can measure wearers' body temperature or trace their heart activity are now available to patients and medical practitioners, and microsized biosensors in a textile patch can help analyze body fluids to provide a good assessment of the subject's health.

Some of the many composite technologies used to manufacture advanced medical and healthcare products are described below.

5.5.1 Polymer blending

A polymer blend or mixture is analogous to metal alloys, in which at least two polymers are combined to create a new material with different physical properties. Polymer blends can be broadly divided into three categories.

1. Immiscible or heterogeneous polymer blends: The constituent polymers exist in separate phases and the respective glass transition temperatures are observed.
2. Compatible polymer blends: This is an immiscible polymer blend that exhibits macroscopically uniform physical properties, caused by sufficiently strong interactions between the component polymers.
3. Miscible or homogeneous polymer blends: These mixtures are often made from polymers with similar chemical structures, resulting in a polymer blend with a single-phase structure. One glass transition temperature is observed.

Thanks to the advancement of nanotechnology, entirely new features can be added to medical textile materials by blending nanosized materials into conventional textile products. For example, silver-containing nanoparticles have been blended into the spinning solutions of alginate and chitosan to prepare alginate and chitosan fibers with antimicrobial properties. When in contact with wound exudate, silver ions are released from these fibers into the exudate and are then absorbed by bacteria and fungi, leading to the death of these undesirable microbes in infected wounds. In the wound management field, polymer blending has also been used to enhance the gelling properties of calcium alginate fibers by mixing sodium alginate and carboxymethyl cellulose (CMC) in an aqueous solution before the wet-spinning process. The CMC component in the composite fiber disrupts the regular structures of calcium alginate, making it easy for the alginate/CMC composite fiber to swell in the presence of saline. As can be seen in Table 5.1, Urgosorb™, which is an alginate/CMC composite fiber containing about 15% CMC, showed a significantly higher saline swelling ratio than Algosteril™, made of calcium alginate. The absorbency of an Urgosorb™ nonwoven dressing is more than 30% higher than that of Algosteril™ under similar test conditions.

Table 5.1 Comparison of alginate/CMC composite dressing to pure alginate dressings

Tests	Urgosorb™	Algosteril™
Absorbency, g/g	20.35	14.27
Gel swelling ratio in water, g/g	4.25	1.85
Gel swelling ratio in saline, g/g	9.65	5.23

5.5.2 Microencapsulation

Microencapsulation is a technology that uses microcapsules to serve as tiny containers of substances, liquid or solid, which are released to fulfill a specific purpose. In a microcapsule, a coating acts as the wall, shell, or membrane to surround the material inside, which can be released in a variety of ways depending on the characteristics of the capsule wall, including physical pressure, friction, diffusion, wall dissolution, and biodegradation. Microencapsulation can be combined with textile products for the application of fragrances, skin softeners, insect repellent, and antimicrobial agents. Examples of microencapsulation include mosquito-repellent finishes applied to outdoor clothing and mite protection applied to mattresses and mattress pads. The insect-repellent finishing agent is encapsulated as a surface finish on the textile and released over time during product use.

Microencapsulation is by far the most frequently used technology to micropackage cosmetic ingredients in textiles, where microcapsules typically have mean diameters of 1–10 μm, as compared to a range of 5–30 μm for a single textile fiber—from about 5 μm for microfibers to about 30 μm for coarse wool. The microcapsules can be modified in size, mechanical robustness, and permeability to customize the release profile to best fit the intended functionality.

Because of their versatility and cost effectiveness, microencapsulation based on urea or melamin-formaldehyde has been the most commonly used technology for cosmetotextiles. The initial step is to emulsify a liquid hydrophobic blend of cosmetic ingredients with the help of a polymer and mechanical homogenization. The polymer is then cross-linked with a reactive urea or melamin-formaldehyde prepolymer, followed by a process to reduce free formaldehyde. Since most of the residual free formaldehyde is removed during the drying/curing steps which take place at the end of the textile finishing process, compliance with the limits of the Oeko-Tex Standard 100, Class 2 (<75 ppm) or even Class 1 (not detectable, usually understood as less than 20 ppm) can be achieved.

Another possible approach is based on cyclodextrins, which are bucket-shaped molecules composed of six to eight glucose units and are enzymatically derived from starch. They have the capability of building protective complexes with a variety of molecules, including cosmetically relevant ones such as menthol and vitamin E. These complexes can be attached to fabrics with the help of an adhesive or a binder, which includes cross-linkable silicones, polyacrylates, polyethylene-vinyl acetate, and polyurethanes. A certain amount of binder, typically 0.25–4.0% of dry matter per weight of fabric, is required to bind the microcapsules, complexes, or loaded particles effectively to the medical textile materials.

5.5.3 Drug carrier and release systems

Drugs and other functional materials in medicine and pharmacy can be incorporated into fibers in a similar way to antimicrobial agents. For example, bovine serum albumin can be loaded into nanofibers in the electrospinning process. Research findings show that the release rate can be strongly controlled by the fiber architecture, and that the enzymes are highly active when released. In addition to biological compounds such as albumin, superparamagnetic nanoparticles such as iron oxide can be incorporated within the nanofibers, nanotubes, and nanorods to guide the nano objects to specific areas within the body.

Electrospinning provides great flexibility in selecting matrix materials for drug delivery. Both degradable and nondegradable polymers can be electrospun into nanofibers and loaded with drugs. Accordingly, a variety of drugs can be encapsulated, including antibiotics, anticancer drugs, proteins, and DNA. The drug-loaded fibrous mats can be used for transdermal drug delivery, wound dressing, and many other applications.

Two commonly used textile finishing processes, the padding process and the exhaust process, can be used to apply active ingredients to textile materials. In the padding process, the fabric is first run through a finishing bath which contains an aqueous dispersion of microcapsules along with the binder, as well as auxiliaries such as softeners or wetting agents. After emerging from the bath, the fabric is squeezed by a pair of rubber rollers with a constant pressure in order to reach a defined wet pick-up level. It then travels through a heat-setting oven for drying and cross-linking of the binder. The fabric stays in the oven for a few seconds to 1–2 min at temperatures of 105–140 °C.

A different principle is used in the exhaust process, where the fabric or garments are first introduced into a closed vessel, such as a dyeing apparatus, with an initial quantity of water. The microcapsule dispersion, the binder, the auxiliaries, and possibly acids/bases for pH adjustment are then metered into the vessel at defined times, speeds, temperatures, and pH in such a way that the microcapsules and the binder ultimately end up distributed as evenly and completely as possible on the textiles. The process is generally a combination of exhaustion and controlled precipitation, which is a sophisticated version of the way softeners work in the domestic washing machine.

The application of microcapsules via padding allows a high degree of variability in the binder-to-microcapsule ratio, while application via the exhaust process requires a very precise steering of temperature and pH—greatly limits the choice of binders, since few of them exhaust well. Other finishing processes such as spraying, printing, and coating play a less important role in the delivery of bioactive ingredients, but they are possible options. For example, the printing and coating processes allow preferential application of microcapsules to one side of the fabric, which is beneficial for the transfer of bioactive ingredients to the skin.

5.5.4 Tissue engineering

Tissue engineering merges the fields of cell biology, material science and engineering, and surgery to fabricate new functional tissue using living cells in a matrix or scaffold. Three-dimensional textile structures made from natural or synthetic materials are widely used as scaffolds in tissue engineering, where viable human tissues are cultured outside the body and then implanted into the body.

In terms of the textile structure used as scaffolds for tissue engineering, it is desirable to produce textile materials that do not alter their mechanical properties when tissue grows in them and forms a composite. The potential of the embroidery technique has been investigated for the development of textile scaffold structures for tissue engineering. Results show that the embroidered textile does not undergo stiffening, whereas in knitted fabric a rise in tensile force is observed. In addition to the textile structure, fiber structure also influences the performance of artificial tissues. For example, polyglycolic acid is widely used for tissue engineering. However, it is rapidly resorbed, which can lead to premature weakening of the tissue before its remodeling. In tissue engineering it is often necessary to improve the mechanical properties of the tissue structure and regulate cell phenotype via interaction with the polymer. In this respect, alginate has specialized uses in textile scaffolds, which may be knitted, woven, nonwoven, braided, embroidered, or combined. The gel-forming ability of alginate is particularly useful when considering the importance of maintaining a stable cell culture condition for different cell types adhered to and proliferated inside the matrix.

Since nonwovens have high porosity, which allows permeation and growth of cells in vitro and in vivo, they are ideally suited for the construction of a comfortable and porous three-dimensional scaffold structure. A nonwoven fibrous mat made of graphite and Teflon has been used around orthopedic implants to promote tissue growth. Resorbable spun-laid nonwovens are used increasingly as a substrate for three-dimensional tissue engineering of chondrocytes and endocrine cells. As a structural mimic of an extracellular matrix, electrospun nanofibers have also been studied as scaffolds for cell growth, and a great number of polymers, including biodegradable polymers such as polylactide, nondegradable polymers such as polyurethane, and natural polymers such as collagen, have been electrospun into porous nonwoven mats to support cell proliferation.

So far, most of the research has been focused on skin, blood vessel, bone, nerve, and cartilage tissue engineering. Attempts to replicate the complicated nature of native extra cellular matrix (ECM) have led to the investigation of a wide range of electrospun nanofibers with tailored biological, chemical, and mechanical properties. The capability to generate porous structures with a large surface area and the versatility to modify bulk and surface properties make electrospun nanofibers excellent candidates for tissue engineering applications.

5.5.5 Bioengineered artificial organs

Nowadays tissue engineering covers a wide range of medical applications. For example, scaffold in the form of a crocheted mesh manufactured from multifilament yarns can be used to grow human dermal fibroblasts. The resulting dermal tissue contains metabolically active cells and a dermal matrix, which provides a viable alternative to autograft skin in various clinical scenarios, from covering raw areas in burn wounds to stimulating the healing of chronic wounds. Such products can be single layered or bi-layered. Due to their superior biocompatibility and mechanical properties, electrospun nanofibers made from silk-fibroin can be used to produce two- and three-dimensional matrices with great potential for tissue regeneration and repair.

In addition to tissue-engineered skins, there have been many attempts to develop bioengineered artificial kidneys which can perform metabolic, endocrine, immune, and other functions in addition to dialysis and ultrafiltration. The artificial kidneys are similar to a conventional hemodialyzer, except that viable cells are introduced to the extracapillary spaces and then attached to the outer surfaces of the hollow fibers. Alternatively, these cells can be entrapped in gels or immobilized on to polymeric microcarriers to avoid blocking the pores or disrupting the mass exchange across the membrane. The viable cells come from many sources, such as human and porcine kidney cells, and after harvesting, culturing, and seeding them on to the hollow fibrous bioreactors, they are expected to perform some of the metabolic functions of a kidney.

Similar to artificial kidneys, hollow-fiber-based artificial liver devices can be developed by culturing living animal or human hepatocytes inside the hollow fibers and circulating the patient's blood in the extracapillary space. Toxic components in the blood that generally diffuse through the hollow fiber membrane are metabolized by the entrapped hepatocytes, and will then either diffuse back into the bloodstream or be washed out by the intraluminal stream.

Hollow fiber bioreactors have also been applied in other extracorporeal devices, such as artificial lungs and blood oxygenators that serve as the cardiopulmonary bypass during an open heart surgery, where the hollow fiber membranes are used to separate blood from gas. Blood flows outside and across bundles of hollow fibers while gas (usually oxygen or oxygen/nitrogen mixtures) flows inside the fibers, and the diffusion of oxygen into and out of the bloodstream is driven by the concentration gradient across the membranes. In a similar way, artificial lungs can also be made by winding a microporous woven fabric around a central tube to form a bundle; blood flow takes place across the woven fabric bundle while gas flows inside, and oxygen diffuses in a way similar to that in a hollow fiber blood oxygenator.

An artificial pancreas can also be developed for the treatment of diabetes. Islets of Langerhans are integrated into a flat sheet or hollow fiber membranes, which separate the cells from the blood and are permeable to glucose and insulin but impermeable to immunoglobulins and lymphocytes.

5.6 Summary

Textile materials can be functionally modified by modifying the polymer, fiber, and textile structures. Over the years many techniques have been developed in the polymer and textile industries, with many related technologies developed in bioengineering, chemical engineering, material processing, and medical sciences. Many unique properties can be obtained by applying chemical, physical, and biological treatments to polymers, fibers, textile materials, and composites, making it possible to enhance the performance of textile materials and develop medical textile products with high absorbency and antimicrobial, antiviolet, drug-releasing, protection, and many other functions that are highly valuable in the field of biomedical materials and functional medical textiles.

Sources for further information

[1] Bartels VT, editor. Handbook of medical textiles. Cambridge: Woodhead Publishing Ltd; 2011.
[2] Pawar SN, Edgar KJ. Alginate derivatization: a review of chemistry, properties and applications. Biomaterials 2012;33:3279–305.
[3] Jayakumar R, Rajkumar M, Freitas H, et al. Bioactive and metal uptake studies of carboxymethyl chitosan-graft-D-glucuronic acid membranes for tissue engineering and environmental applications. Int J Biol Macromol 2009;45(2):135–9.
[4] Sun S, Wang A. Adsorption kinetics of Cu(II) ions using N,O-carboxymethyl-chitosan. J Hazard Mater 2006;131(1–3):103–11.
[5] Gabbay J, Mishal J, Magen E, et al. Copper oxide impregnated textiles with potent biocidal activities. J Ind Text 2006;35:323–35.
[6] Borkow G, Gabbay J. Biocidal textiles can help fight nosocomial infections. Med Hypotheses 2008;70(5):990–4.
[7] Hipler UC, Elsner P, Fluhr JW. Antifungal and antibacterial properties of a silver-loaded cellulosic fiber. J Biomed Mater Res 2006;77(1):156–63.
[8] Costain DJ, Kennedy R, Ciona C, et al. Prevention of postsurgical adhesions with N,O-carboxymethyl chitosan: examination of the most efficacious preparation and the effect of N,O-carboxymethyl chitosan on postsurgical healing. Surgery 1997;121(3):314–9.
[9] Chen RN, Wang GM, Chen CH, et al. Development of N,O-(carboxymethyl)chitosan/collagen matrixes as a wound dressing. Biomacromolecules 2006;7(4):1058–64.
[10] Janvikul W, Uppanan P, Thavornyutikarn B, et al. Fibroblast interaction with carboxymethylchitosan-based hydrogels. J Mater Sci Mater Med 2007;18(5):943–9.
[11] Sather HV, Holme HK, Maurstad G, et al. Polyelectrolyte complex formation using alginate and chitosan. Carbohydr Polym 2008;74:813–21.
[12] Li X, Xie H, Lin J, et al. Characterization and biodegradation of chitosan-alginate polyelectrolyte complexes. Polym Degrad Stab 2009;94:1–6.
[13] Shao X, Hunter CJ. Developing an alginate/chitosan hybrid fiber scaffold for annulus fibrosus cells. J Biomed Mater Res Part A 2007;82:701–10.
[14] Wan ACA, Liao IC, Yim EKF, et al. Mechanism of fiber formation by interfacial polyelectrolyte complexation. Macromolecules 2004;37:7019–25.
[15] Tamura H, Tsuruta Y, Tokura S. Preparation of chitosan-coated alginate filament. Mater Sci Eng C 2002;20:143–7.
[16] Knill CJ, Kennedy JF, Mistry J, et al. Alginate fibres modified with unhydrolysed and hydrolysed chitosans for wound dressings. Carbohydr Polym 2004;55(1):65–76.
[17] Buchenska J. Polyamide fibers with antibacterial properties. J Appl Polym Sci 1996; 61:567–76.
[18] Chilarski A, Krucinska I, Kiekens P, et al. Novel dressing materials accelerating wound healing made from dibutyrylchitin. Fibres Text East Eur 2007;15:77–81.
[19] Ciechanska D. Multifunctional bacterial cellulose/chitosan composite materials for medical applications. Fibres Text East Eur 2004;12:69–72.
[20] Eaglstein WH. Moist wound healing with occlusive dressing: a clinical focus. Dermatol Surg 2001;27:175–81.
[21] Ehret P, Lahteenkorva K. Nonwoven/film biodegradable composite structure. US Patent 5783504, 1998.
[22] Qin Y. Silver containing alginate fibres and dressings. Int Wound J 2005;2(2):172–6.
[23] Graham K, Schreuder-Gibson H, Gogins M. Incorporation of electrospun nanofibers into functional structures. INJ 2004;13:21–7.

[24] Grechin AG, Buschmann HJ, Schollmeyer E. Quantification of cyclodextrins fixed onto cellulose fibers. Text Res J 2007;77:161–4.
[25] Gulrajani ML. Nano finishes. Indian J Fibre Text Res 2000;31:187–201.
[26] Petrulyte S. Advanced textile materials and biopolymers in wound management. Danish Med Bull 2008;55(1):72–7.
[27] Baumgarten P. Electrostatic spinning of acrylic microfibers. J Colloid Interface Sci 1971;36:71–6.
[28] Larrondo L, Manley RS. Electrostatic fiber spinning from polymer melts 1. Experimental observations on fiber formation and properties. J Polym Sci Part B: Polym Phys 1981;19:909–15.
[29] Rajendran S, Anand SC. Contribution of textiles to medical and healthcare products and developing innovative medical devices. Indian J Fibre Text Res 2006;31:215–29.
[30] Blackburn RS, editor. Biodegradable and sustainable fibres. Cambridge: Woodhead Publishing Ltd; 2005.
[31] Qin Y, Hu H, Luo A, et al. The effect of carboxymethylation on the absorption and chelating properties of chitosan fibers. J Appl Polym Sci 2006;99(6):3110–5.
[32] Anand SC, editor. Medical textiles 96. Cambridge: Woodhead Publishing Ltd; 1997.
[33] Rigby AJ, Anand SC, Horrocks AR. Textile materials for medical and healthcare applications. J Text Inst 1997;88(Part 3):83–93.
[34] Tamura H, Tsuruta Y, Tokura S. Preparation of chitosan-coated alginate filament. Mater Sci Eng 2002;20(1-2):143–7.
[35] Wang J, Smith J, Babidge W, et al. Silver dressings versus other dressings for chronic wounds in a community care setting. J Wound Care 2007;16:352–6.
[36] Min BM, Lee G, Kim SH, et al. Electrospinning of silk fibroin nanofibers and its effect on the adhesion and spreading of normal human keratinocytes and fibroblasts in vitro. Biomaterials 2004;25:1289–97.
[37] Lee HJ, Yeo SY, Jeong SH. Antibacterial effect of nanosized silver colloidal solution on textile fabrics. J Mater Sci 2003;38:2199–204.
[38] Qin Y, Gilding DK. Fibres of co-spun alginates. US Patent 6080420, 2000.
[39] Mathiowitz E, editor. Encyclopaedia of controlled drug delivery. New York: Wiley Interscience; 1999.
[40] Patrick CW, Mikos AG, McIntire LV, editors. Frontiers in tissue engineering. Oxford: Pergamon Press; 1998.
[41] Shonaike GO, Advani SG, editors. Advanced polymeric biomaterials. Boca Raton: CRC Press; 2003.
[42] Chasin M, Langer R, editors. Biodegradable polymers as drug delivery systems. New York: Marcel Dekker; 1990.

Superabsorbent polymers and their medical applications

6

6.1 Introduction

Superabsorbent polymers (SAPs) are polymers that can absorb and retain extremely large amounts of liquid relative to their own mass. In deionized or distilled water a SAP may absorb 50,000% its own weight of liquid, while in a 0.9% saline solution the absorbency can reach about 5000% its own weight. This compares to water absorbencies of 180%, 400%, 1050%, 1200%, and 1890% for the filter paper, facial tissue paper, soft polyurethane sponge, wood pulp fluff, and cotton balls respectively.

SAPs are usually made of cross-linked hydrophilic polymers containing anionic water-holding groups such as carboxylic acid groups. Common SAPs are generally white sugar-like hygroscopic materials, and are divided into two main classes: synthetic (petrochemical based) and natural (polysaccharide and polypeptide based). For commercial reasons, most current SAPs are produced from acrylic acid (AA), its salts, and acrylamide via solution or inverse-suspension polymerization techniques. The total absorbency and swelling capacity is controlled by the type and degree of cross-linkers used to make the polymer. Low-density cross-linked SAPs generally have a higher absorption capacity and swell to a larger degree, while polymers with a high cross-link density exhibit a lower absorption capacity and swell to a lesser extent, but the resultant gel is firmer and can maintain the particle shape even under modest pressure. The high absorption capacity of SAPs makes them ideal materials for use in water-absorbing applications such as diapers for children and adults, feminine hygiene products, and other similar applications, where the superabsorbent materials have advantages over traditional absorbent materials such as cloth, paper wadding, and cellulosic fibers.

Since the first commercial use of SAPs in feminine napkins in Japan in 1978, a large number of end uses have been developed. The greatest use of SAPs is found in personal disposable hygiene products, such as baby diapers, adult protective underwear, and sanitary napkins. In addition, SAPs are suited in areas such as contact lenses, drug administration systems, reconstruction of human muscle cells and blood vessels, wound dressings, and many other novel end uses. The global use of SAPs has exceeded 1 million tons, with baby diapers, adult incontinence products, feminine hygiene products, and other types of healthcare products accounting for 81%, 8%, 5%, and 6% respectively.

Medical Textile Materials. http://dx.doi.org/10.1016/B978-0-08-100618-4.00006-6

6.2 Development of superabsorbent polymers

The synthesis of the first water-absorbent polymer goes back to 1938, when AA and divinylbenzene were thermally polymerized in an aqueous medium. In the late 1950s the first generation of hydrogels appeared, mainly based on hydroxyalkyl methacrylate and related monomers, and with swelling capacity up to 40–50%. These hydrogels were used in the development of contact lenses that revolutionized ophthalmology.

The first commercial SAP was produced through alkaline hydrolysis of starch-graft-polyacrylonitrile, and was developed in the 1970s at the Northern Regional Research Laboratory of the US Department of Agriculture. Expense and inherent structural disadvantages of this product, such as a lack of sufficient gel strength, are seen as the major factors in its early market defeat.

Commercial production of SAPs began in Japan in 1978 for use in feminine napkins, and further developments led to SAP materials being used in baby diapers in Germany and France in 1980. In 1983 low-fluff diapers containing 4–5 g SAPs were marketed in Japan, followed shortly by the introduction of thinner superabsorbent diapers in other Asian countries, the US, and the Europe. By 1990 the world production of SAP resins exceeded 1 million tons.

According to the European Disposables and Nonwovens Association, the total production of SAPs in 2005 approached 1,483,000 tons, with 623,000 tons reported in Asia (mostly by Nippon Shokubai, San-Dia Polymers, and Sumitomo Seika Chemicals), 490,000 tons in North America (by Degussa, BASF, Dow, and Nippon Shokubai), and 370,000 tons in Europe (mostly by Degussa and BASF). The SAPs are mainly made from partially neutralized, lightly cross-linked polyacrylic acid, which has been proven to give the best performance versus cost ratio. The polymers are synthesized, dried, and milled into granular white solids. When placed in water they swell to a rubbery gel that in some cases can contain up to 99 wt% water.

6.3 Structural characteristics of superabsorbent polymers

Because of their ability to swell in the presence of water, SAPs are also known as hydrogels. Based on their structural features, SAPs are divided into two groups. The first is physical gels, also called reversible gels, which have a three-dimensional space network where the polymer chains are bonded by electrostatic force, hydrogen bond, and hydrophobic interactions. The second group is chemical gels, which are stable gels with a three-dimensional space network in which the polymer chains are permanently covalently bonded. Typical physical gels include gelatin and agar-agar, which are unstable at high temperatures, while chemical gels include mostly cross-linked SAPs.

In order to swell in water, the chemical structures of the SAPs should be controlled by controlling the type of polymer matrix, characteristics of functional groups, degree of cross-linking, quantity of initiator, reaction temperature, chemical composition, and other important factors. Small amounts of cross-linkers play a major role in modifying the properties of SAPs.

In the presence of aqueous fluid SAPs undergo a process of swelling, in which water is absorbed into the polymer structure. The final swelling capacity, or centrifuge retention capacity, is the amount of 0.9 wt% saline solution that a SAP can retain under free swelling conditions when surface water has been removed in a centrifuge. During the absorption process, there is a balance between the forces of retraction and the tendency for the chains to swell to infinite dilution. The degree of cross-linking has a direct effect on the level of swelling of the polymer and the strength of the network. Increased cross-linking density results in decreased swelling capacity and increased gel strength.

In ionic polymers, the chains that contain charges repel each other. Overall electrical neutrality is maintained, as the negative carboxylate groups are balanced by the positive sodium ions. Upon contact with water the sodium ions are hydrated, which reduces their attraction towards the carboxylate ions due to the high dielectric constant of water. This allows the sodium ions to move freely within the network, which contributes to the osmotic pressure within the gel. However, the mobile positive sodium ions cannot leave the gel because they are still weakly attracted to the negative carboxylate ions along the polymer backbone and behave as if they are trapped by a semipermeable membrane. Since the driving force for swelling is the difference between the osmotic pressure inside and outside the gel, increasing the level of sodium ions outside the gel will lower the osmotic pressure and reduce the swelling capacity of the gel. The maximum swelling of the gel will occur in deionized water.

During the swelling process, water diffuses into a particle of a SAP when the concentration of water is initially lower in the interior of the particle. As water travels into a particle, it swells to accommodate the additional molecules. If the SAP is more highly cross-linked, it is more rigid in the swollen state. By improving the rigidity of the particles it is possible to enable the swelling particles to push aside the surrounding materials, thereby maintaining porosity and permeability during subsequent contacts with liquid.

6.4 Classification of superabsorbent polymers

SAPs can be classified into four groups on the basis of the presence or absence of electrical charge located in the cross-linked chains.

1. Nonionic.
2. Ionic, including anionic and cationic.
3. Amphoteric electrolyte containing both acidic and basic groups.
4. Zwitterionic containing both anionic and cationic groups in each structural repeating unit, such as polybetaines.

SAPs are also classified based on the type of monomeric unit used in their chemical structure. Most conventional SAPs fall into one of the following categories.

- Cross-linked polyacrylates and polyacrylamides.
- Hydrolyzed cellulose-polyacrylonitrile (PAN) or starch-PAN graft copolymers.
- Cross-linked copolymers of maleic anhydride.

SAPs can also be divided into synthetic and natural types, with the latter being further divided into two main groups: those based on polysaccharides and those based on polypeptides. Natural SAPs are usually prepared by the addition of some synthetic parts to the original substrates, for example, graft copolymerization of vinyl monomers on polysaccharides. Because of their low price and high performance, synthetic SAPs are widely used in many healthcare and hygiene products, and the term "superabsorbent polymers" usually implies the most conventional type of SAP, that is, an anionic copolymeric network based on partially neutralized AA or acrylamide.

In terms of their physical forms, SAPs can be made into powders, fibers, sheets, granular particles, etc. The most commonly available SAPs are hard, dry, granular powders that look like clean white sand or granular sugar, and are widely used for enhancing the absorbency and moisture-retention capacities of medical and hygiene products. Compared to powders, synthetic absorbents in a fiber form can be more easily introduced into textile structures. In particular, when combined with other types of fibers they remain in place during handling and even after swelling.

6.5 Synthesis of superabsorbent polymers

A variety of monomers, mostly acrylics, is used to synthesize SAPs. Acrylic acid, and its sodium or potassium salts, and acrylamide are most often used in the industrial production of SAPs. The AA monomer is inhibited by methoxyhydroquinone to prevent spontaneous polymerization during storage. A number of other monomers such as methacrylic acid, methacrylamide, acrylonitrile, 2-hydroxyethylmethacrylate (HEMA), *N*-vinyl pyrrolidone, vinyl sulphonic acid, and vinyl acetate, and biopolymers such as cellulose, starch, chitosan, gelatin, and some of their derivatives are also used as raw materials for the production of SAPs.

Figure 6.1 shows the chemical structure of polyacrylic acid. During the polymerization process for acrylic-based SAPs, water-soluble thermal initiators such as potassium persulfate and ammonium persulfate are often used in solution and inverse-suspension methods of polymerization. Redox pair initiators such as Fe^{2+}-H_2O_2 (Fenton reagent) are also employed. The production process uses one of two general pathways to prepare the SAP networks, that is, simultaneous polymerization and cross-linking by a polyvinylic cross-linker, and cross-linking of a water-soluble prepolymer by a polyfunctional cross-linker.

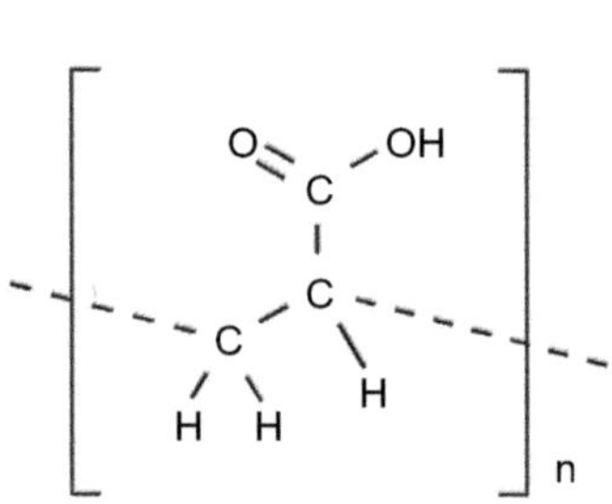

Figure 6.1 The chemical structure of polyacrylic acid.

Several parameters are important for the production of high-quality SAPs. First, a good balance must be maintained between the molecular weight of the polymer and the yield of the production process. Low-molecular-weight species are extractable, whereas high-molecular-weight species require a longer polymerization time. Commercial SAPs typically contain 5–20% extractables.

Cross-linking is another important part of the SAP production process. SAPs are often covalently cross-linked by using cross-linkers containing two or more polymerizable double bonds, which are incorporated into the backbone of the polymer chains as they grow during the polymerization reaction. The cross-linking can occur either in the bulk of the SAPs, which normally takes place during the polymerization stage of superabsorbent production, or on the surface of the SAPs, which is a newer process that improves the absorption-against-pressure profile of the polymer. Cross-linking during polymerization is important, in that it yields a networked polymer which will not dissolve in water and can absorb and retain water under low load. A typical cross-linking agent is trimethylolpropane triacrylate at concentrations of 0.05 mol% relative to the monomer. Cross-linking is also possible with ethylene glycol diglycidyl ether, which reacts with carboxyl groups on the polymer molecules.

Cross-linking the surface of SAP particles can help improve the absorption against pressure and the swelling rate of a SAP. This type of post-cross-linking is normally performed on the dried, milled, and sized SAP as the final stage of the production process, when a cross-linking solution is applied to the particles and then "cured" by heating. The result of this process is an increase in the density of cross-linking on the surface of the particles, giving what can be described as a core-shell particle where the core is lightly cross-linked while the shell has a higher cross-linking density. The cross-linking process is initiated with gamma radiation emitted by cobalt-60 or caesium-137 and also an accelerated electron stream.

In summary, the following reaction variables affect the properties of acrylic anionic SAPs.

1. Cross-linker type and concentration.
2. Initiator type and concentration.
3. Monomer type and concentration.
4. Type, size, and amount of inorganic particles incorporated.
5. Polymerization method.
6. Polymerization temperature.
7. Amount and type of surfactant used.
8. Stirrer/reactor geometry and rate of stirring.
9. Porosity generating method or the amount and type of the porogen.
10. Drying method, temperature, and time.
11. Post-treatments such as surface cross-linking to enhance the swollen gel strength.

6.6 Characterization of superabsorbent polymers

The main property of a SAP is its ability to hold water. In addition to this, other functional features of an ideal SAP material can be listed as follows.

1. The highest absorption capacity or maximum equilibrium swelling in saline.
2. Desired rate of absorption depending on the application requirement.
3. The highest absorbency under load (AUL).
4. The lowest soluble content and residual monomer.
5. The lowest price.
6. The highest durability and stability in the swelling environment and during storage.
7. The highest biodegradability without formation of toxic species following degradation.
8. pH-neutrality after swelling in water.
9. Colorless, odorless, and nontoxic.
10. Photostability.
11. Rewetting capability (if required).

Depending on the application requirement, for example, in hygienic or agricultural applications, SAPs should be able to hold the imbibed solution or donate it. Obviously, it is not possible for a SAP sample simultaneously to fulfill all the above-mentioned features. In fact, the synthetic components for achieving the maximum level of some of these features will lead to inefficiency in the rest. Therefore, in practice the reaction variables in the production process must be optimized such that an appropriate balance between the properties is achieved.

According to the European Disposables and Nonwoven Association, the tests for polyacrylate SAPs include pH, residual monomers, particle size distribution, moisture content, free swelling capacity, centrifuge retention capacity, absorption under pressure, flowrate, density, extractables, respirable particles, and dust.

6.7 Absorption properties of superabsorbent polymers

Chemically, SAPs are made of polymer chains linked together at selected points, due to which they form a three-dimensional cross-linked structure. In the dry state, the polymer chains take the form of tightly folded coils. Upon wetting, the functional groups on the polymer chains undergo salvation and dissociate under the influence of water, when positive ions and negative charges bonded to polymer chains repel each other under the influence of electrostatic forces. This leads to the loosening of the polymer coil, thus enabling water to penetrate into the polymer to form a gel. The process ends when the polymer chains that form a spatial network have extended to the maximum, resulting in the absorption of a large amount of water into the SAP.

Figure 6.2 shows a SAP sample before and after wetting. As water is absorbed the SAP swells in volume, and hence the absorption of SAPs is also called swelling, often expressed as grams of solution or water taken up by 1 g of dry gel. The absorption capacity can amount from a few to a few hundred grams of solute per gram of superabsorbent, and depends on many factors. The type of functional groups and the degree of polymer cross-linking have a decisive influence on this phenomenon, while the composition of the absorbed solution is also an important factor. The presence of electrolytes and chemical compounds capable of reacting with the functional groups in the gel usually decreases the absorption capacity of the SAPs. While a typical SAP can absorb 500 times its own weight (30–60 times its volume) of distilled or deionized

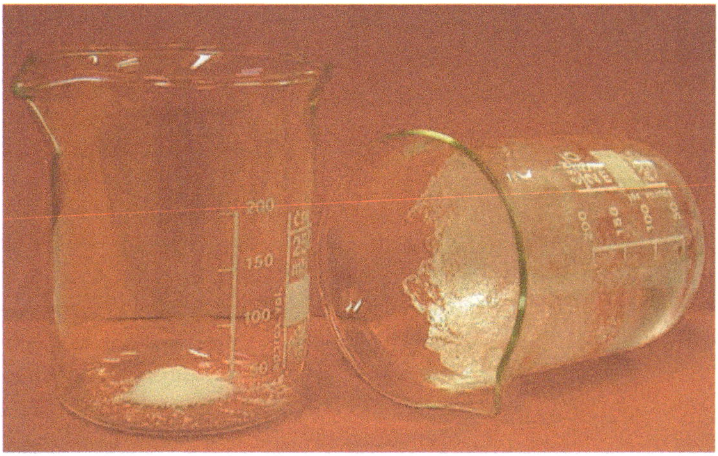

Figure 6.2 An illustration of a SAP sample before and after wetting.

water, it can only absorb about 50 times its own weight of a 0.9% saline solution. The total absorbency and swelling capacity are controlled by the type and the degree of cross-linking in the polymers. A lower-density cross-linked polymer has a higher absorption capacity, and a softer and more cohesive gel can be formed from it.

During the absorption process, the diffusion of water into the SAP is controlled by the affinity between the functional groups in the SAP and the molecules in the contacting fluid, where the swelling equilibrium is determined by the balance of three main forces.

1. Free energy of the mixture of network chains and solvent.
2. Osmotic pressure inside the network resulting from the mobility of counter ions surrounded by the groups of permanent charge.
3. Flexible network response or flexible swelling pressure.

For SAPs with anionic groups on the main chain, the first two forces favor swelling of gel, in contrast to force three.

6.8 Test methods for the evaluation of superabsorbent polymers

Because of their wide range of applications, the properties of SAPs have been evaluated by academic researchers and industrial analysts in many different ways. Some of the test methods are described below.

6.8.1 Free absorbency capacity

Generally, when the terms of swelling or absorbency are used without specifying its conditions, they imply uptake of distilled water while the sample is freely swollen, that is, no load is put on the testing sample. The free absorbency capacity can be tested by the following methods.

Teabag method: This method is the most conventional and fastest way to assess the absorption capacity of a SAP sample, suitable for limited amounts of samples (W_0=0.1–0.3 g). The SAP sample is placed into a teabag and the bag is dipped in an excess amount of water or saline solution for 1 h to reach equilibrium swelling. Then excess solution is removed by hanging the bag until no liquid drops off. The teabag is weighed and the swelling capacity = $(W_1 - W_0)/W_0$, where W_1 is the weight of the swollen sample and W_0 is the original weight of the sample.

Centrifuge method: Centrifugal data are more accurate than the teabag method and are occasionally reported in patents and data sheets. In this method, about 0.2 g (W_1) of SAP is placed into a bag (60×60 mm) made of nonwoven fabric. The bag is dipped in 100 mL of saline solution for half an hour at room temperature. It is taken out and excess solution is removed with a centrifugal separator for 3 min. Then the weight of bag (W_2) is measured. The same stages are carried out with an empty bag, and the weight of bag (W_0) is measured. The swelling capacity = $(W_2 - W_0 - W_1)/W_1$. Since the interparticle liquid is noticeably removed by this method, the measured values are often more accurate and lower than those obtained from the teabag method.

6.8.2 Absorbency under load

The AUL data are usually given in the patent literature and technical data sheets by industrial SAP manufacturers. When the term AUL is used without specifying the swelling media, it implies an uptake of 0.9% NaCl solution while the testing sample is pressurized by a load, often specified to be pressure of 0.3, 0.6, or 0.9 psi. A typical AUL tester is a simple but finely made device including a macroporous sintered glass filter plate placed in a petri dish. The weighed dried SAP sample (0.90 ± 0.01 g) is uniformly placed on a surface of polyester gauze located on the sintered glass. A cylindrical solid load is put on the dry SAP particles while it can be freely slipped in a glass cylinder. After the desired load is placed on the SAP sample, saline solution (0.9% NaCl) is added into the petri dish to keep the liquid level equal to the height of the sintered glass filter. The whole set is covered to prevent surface evaporation and probable change in the saline concentration. After 1 h the swollen particles are weighed again, and AUL = $(W_2 - W_1)/W_1$, where W_1 and W_2 denote the weight of dry and swollen SAP, respectively.

6.8.3 Swelling rate

The vortex method is often employed in research and technical laboratories to offer a quick assessment of SAP swelling rate. Water or saline solution (50.0 g) is poured in a 100 mL beaker and its temperature is adjusted at 30 °C. It is stirred at 600 rpm using a magnetic stirrer. The SAP sample (mesh 50–60, W_0=0.50–2.0 g) is added and a stopwatch is started. The time elapsing from the addition of the SAP into the fluid to the disappearance of vortex (tvd, s) is measured. The swelling rate (SR, g/g.s) = $(50/W_0)$/tvd.

6.8.4 Soluble fraction

The soluble fraction is the sum of all water-soluble species, including noncross-linked oligomers and nonreacted starting materials such as residual monomers. The soluble content is simply measured by extraction of the SAP sample in distilled water. A certain amount of the SAP sample (e.g., 0.10 g) is poured into an excess amount of water and dispersed with mild magnetic stirring to reach equilibrium swelling, before the swollen sample is filtered and oven dried. The sample weight loss during this process represents the soluble fraction.

6.8.5 Residual monomer

High-performance liquid chromatography is often seen as a preferred method to quantify the residual monomer in SAP samples. Orthophosphoric acid solution is usually used as an extractant to remove acrylic acid or salt from the hydrogel network. The acrylic salt is converted to AA at the acidic pH of both the extracting and the eluting media. The separation is usually performed in isocratic mode at a 1.8 mL/min flow rate and ambient temperature on an analytical column, where the mobile phase is an aqueous 0.01% orthophosphoric acid. The UV–vis absorbance over the 190–400 nm range is registered, and the wavelength used for quantification is 200 nm.

6.8.6 Ionic sensitivity

To achieve a comparative measure of sensitivity of the SAP materials to the composition of the aqueous fluid, a dimensionless swelling factor, f, is defined as f = 1 − (absorption in a given fluid/absorption in distilled water). A large f value means a high absorbency loss of the sample due to the presence of salt in the solution. Therefore, SAPs with lower f values are usually preferred.

6.8.7 Test for absorption time

Absorption time is the time needed to absorb a certain amount of fluid, which is strongly correlated to the probability of leakage. In this test, a diaper is laid with the backsheet side on a polyurethane-foam base and covered with a cover plate. Weights are placed on the cover plate to simulate the weight of a baby lying on the diaper. Because of the foam base, pressure is distributed equally across the diaper. The cover plate has a hole which is placed over the area where the diaper typically absorbs urine. An application tube containing an electrode connected to a computer is mounted over that hole. A set amount of synthetic urine is pumped into the application tube, and with the help of the electrode the time is measured until all liquid in the tube is absorbed by the diaper. This is repeated three to four times to simulate the repeated urination of a baby. Results are recorded as absorption rates, that is, volume per time (mL/s).

6.8.8 Test for rewet

A rewet test is used to assess the amount of liquid that has been absorbed but is squeezed out again under pressure, for example, when the wearer sits down. Low rewet is important to keep the wearer's skin dry. In this test a diaper is loaded with a certain amount of liquid, and after a wait (typically 5–10 min) an absorbent material is laid on the topsheet in the area where urine is usually absorbed in real use. The material is weighed before it is applied to the wet diaper. Often filter paper is used as absorbent material, but modern methods use collagen instead, which has absorbent properties closer to those of human skin. Then a weight is applied to the material, simulating the wearer's weight when sitting down. After the weight is removed, the amount of liquid absorbed by the absorbent material is determined by weighing the material again and calculating the difference. Results are reported in weight, either grams or milligrams.

6.8.9 Consumer testing of absorbent hygiene products

Consumer testing is an important methodology in the area of absorbent products, mainly because these products are usually used by consumers rather than professionals. Consumers expect not only good technical performance, such as good leakage protection, but also properties like good wearing comfort and fit, which cannot be tested sufficiently in the laboratory. In-use testing can adequately compare different products and provide guidance for product improvements.

In practice, different consumer-testing techniques are used depending on the aims of the research. Small base testing in 10–30 product users is often used to determine a specific product property, such as the right dimension to deliver a good fit. In this case, different product types are compared by using the same panel list. In a typical study, users in the panel list will put on the product before measurements are taken, directly after application and after a certain wear time. After wear, the product fit will be checked for signs of slippage and the skin will be examined for any evidence of pressure marks.

Interviews or focus-group discussions after product usage are used to assess how a product was perceived by the wearers or, in the case of baby diapers, by the caregivers. The number of consumers interviewed can vary, but is typically at least 20–30 people.

Diary studies are used to assess user experience under realistic in-use conditions. They will typically be conducted with 100 or more users, who complete diary entries after each product usage. Studies usually run for one to two weeks, or, in the case of feminine hygiene products, throughout one menstruation period. Typical questions in a diary would be if a product had leaked; where it had leaked; how long it was used, etc. The users may also be asked to return used products for laboratory assessment, so that the technical measurements of the used product can be compared with its in-use performance as recorded in the diary.

Consumer perception tests might be used to determine perceptions of a product. They are typically given to 100 or more consumers, who are provided with enough products to use for one to two weeks. Consumers are asked to fill in a questionnaire after product usage, in which they are asked about their perception of the product, for example, "how do you rate the product quality overall?". In perception tests, different products might be placed with the same consumer group (paired comparison test) or different sets of consumers (parallel test), depending on the research objectives.

6.9 Applications of superabsorbent polymers in medicine and healthcare

Since their introduction into the consumer market in the 1980s, SAPs have found widespread uses in personal disposable hygiene products, such as baby diapers, adult protective underwear, and sanitary napkins. In addition, these materials are used in a number of novel biomedical products such as surgical caps and drapes, as well as niche products like contact lenses and wound dressings, utilizing their ability to hold a large amount of water in the swollen polymeric structure.

The high absorption capacities of SAPs are particularly suited to absorbent products in personal healthcare and hygiene. These products all contain an absorbent core, which may be made up of natural or synthetic fibers, SAPs, or a mixture of fluff and SAPs. A top layer is used to carry fluid into the absorbent core while maintaining dryness against the skin. This so-called topsheet, which is made of synthetic or natural fibers, and a backsheet, which is usually a polyethylene film or a film/nonwoven composite, are combined with the absorbent core to form a final product that may be used in a number of healthcare and hygienic applications.

6.9.1 Baby diapers

A diaper (in American and Canadian English) or nappy (in Commonwealth English) is a kind of underwear that allows one to defecate or urinate in a discreet manner. Diapers have been worn throughout human history, and are made of cloth or disposable materials. Cloth diapers are composed of layers of fabric made from cotton, hemp, bamboo, or microfibers, and can be washed and reused multiple times. Disposable diapers contain absorbent materials and are thrown away after use. The decision to use cloth or disposable diapers is a controversial one, due to issues ranging from convenience, health, and cost due to their effect on the environment.

Compared to cloth diapers, disposable diapers have superior urine absorbency, leak protection, dryness, and rash-protection properties. Since modern disposable baby diapers were first introduced in the 1940s and mass produced in the 1960s, the demand for these products has expanded rapidly as they became more affordable for everyday use. Since the 1990s a steady stream of technical innovations have further improved the affordability, function, and comfort of diaper products, enabling disposable diapers to be widely used in developing as well as developed nations. The use of SAPs has revolutionized the diaper industry as manufacturers took advantage of the amazing liquid retention ability of these polymers. With refined understanding of the impact of SAPs on the absorbent core, the amount of cellulose pulp fluff used in the diapers was reduced by half, making it possible to produce ultraslim diapers.

During the production process, SAPs in granular or powder forms are added to baby diapers in two ways: layered or blended. In the layered method, powdered SAPs are first scattered on a layer of fluff pulp before it is folded, so the SAPs are located in a centralized layer in the absorbent structure, which is further covered with a nonwoven fabric layer. In the blended application, the SAPs are first mixed homogeneously with the fluff pulp before the mixture is laid down to form the absorbent structure, which is subsequently covered with a nonwoven fabric.

During the development of diapers with improved performance, a separate layer of nonwoven fibers was added to improve urine distribution in the diaper. This distribution layer is placed between the composite absorbent core, consisting of cellulose fiber and SAPs, and the porous cover sheet. The distribution layer has a lower absorbency than either the standard cellulose fluff or SAP, thus allowing fast liquid distribution within the diaper. This layer is often made from nonabsorbent nonwoven materials such as polypropylene fibers, with sufficient porosity to allow liquid to pass through freely.

The diaper core is responsible for holding the liquid during the actual use of the product, and many aspects must be considered during its design—for example, the absorption rate of the diaper must not be slower than the urination rate of the baby, otherwise leakage will occur. The absorption rate of the diaper's composite structure is influenced by the absorption rate of the SAPs. Fast swelling of the polymer may or may not be desirable. In some diaper designs, fast swelling may cause the diaper to leak if the porosity and permeability of the composite are reduced, hindering the penetration of the liquid into the inner part of the diaper. Particle size and shape can affect the maximum absorption capacity of the polymer, and the placement of fast- or slow-absorbing polymers in the composite structure has important implications for the effectiveness of the composites. The particle size, placement, and relative amounts should be adjusted to achieve optimized absorption.

AUL and stability of the gel against shear are important properties of SAPS, and relate strongly to diaper performance. Diaper leakage is closely correlated with the stability of a gel to shearing. Rigid superabsorbent particles created by increasing the cross-linking can have a high gel modulus and help the particles withstand shearing from the baby's weight.

Although baby diapers are used for a relatively short period during the human lifespan, this period of use includes a wide range of developmental progress, from tiny premature babies to toddlers undergoing toilet training. Thus diapers for different stages of a baby's development are excellent examples to showcase the application of different technologies used in absorbent hygiene products. To illustrate the effect of these stages on product requirements, the differences between products made for three stages are described below.

Newborn stage: In the first three to five months after birth, parents are especially concerned with protecting their newborn baby's skin and making the transition from the mother's womb into the world as smooth and comfortable as possible. Babies in this phase experience the world mainly through feeling and listening, while the visual sense is not yet well developed. Thus, in the diaper context, the tactile properties are particularly important. Diapers for newborns are usually made from especially soft materials, and often feature specific technologies to offer good leakage protection. As for any stage, absorption of urine to provide skin dryness and urine leakage protection is important, too.

Determined explorer stage: In this stage babies have become able to walk and are very mobile, so it is important that this mobility is not inhibited by diapers. Diapers for babies in this stage feature not only good absorbency but also elastic elements such as stretch sides to adapt to the baby's body during movement, and are made especially thin and narrow between the legs for a good and comfortable fit.

Capable learner stage: Once babies enter the "capable learner" stage, they start to become more independent from the parents and more self-conscious. They try to do many tasks on their own that parents have previously done for them, such as trying to dress themselves or taking off/putting on diapers by themselves. Putting on a classical taped diaper without help is very hard even for a grown-up, and therefore diapers that are shaped like a pant and can be pulled up are the right choice for babies in this stage.

6.9.2 Feminine hygiene products

Disposable absorbent products for feminine hygiene during menstruation are an important part of life for millions of women. During her lifetime the average woman will have approximately 13 cycles each year for about 37 years, and will spend more than 6 years of that period menstruating. Menstrual flow varies with age, number of children, and use of hormones or intrauterine devices, and typically is heaviest during the second and third days of each period. Many types of products, such as tampons, panty liners, and pads, are used during this period.

In this particular field a number of consumer needs have been identified, including the needs to feel clean and dry, to be protected and in control, and for comfort and convenience. Key product performance requirements include the ability to handle a range of menstrual flows under a variety of use conditions and activities, to be friendly to skin and comfortable, and to be easy to use and dispose of. The most advanced pads for menstrual protection make use of unique topsheets that funnel fluid into the absorbent core, hence maintaining a clean and dry feeling. The core of SAPs rapidly absorbs moisture and locks it away, thus minimizing leaks and controlling odor. Odor control can be done via different routes, including slowing down odor generation by absorbing liquid and transferring it into a gel, using odor-absorbent materials such as zeolite or the like, inhibiting bacteria growth with antimicrobial agents, etc. A backsheet prevents moisture in the core from exiting the pad and staining clothing. Backsheets may be completely impermeable or, as is the case in most modern backsheets, semipermeable ("breathable"), allowing water vapor but not liquid water to pass. This property contributes to comfort.

In recent years disposable absorbent feminine hygiene products have largely replaced older methods of managing menses in women. These products include internal tampons, full-sized sanitary napkins or towels, and panty shields, which may be used to protect undergarments from light menstrual flow or vaginal discharge to maintain cleanliness. In terms of their physical design and basic components, the most advanced sanitary napkins and panty shields make use of many of the same technologies developed for use in baby diapers, including SAPs and nonwoven topsheets that keep skin feeling dry. The requirements for all these products are that they absorb and retain fluid, stay in place, prevent odor, and are comfortable to wear. In Western Europe sanitary towels make up approximately 45% of feminine hygiene products sold, while panty shields make up about 37% and tampons about 18%.

6.9.3 Incontinence products

Single-use disposable incontinence products absorb and lock away urine and feces so that leakage does not occur and the user's skin is kept dry and healthy. These products are available in both diaper and sheet forms, with the latter used as bedding materials. Disposable diapers are composites typically consist of an inner covering layer (coverstock), an absorbent layer, and an outer layer, where coverstock is a generic name used to describe the several nonwoven materials of about 18–85 g/m^2 used in hygiene products, which typically include topsheet, barrier leg cuff, backsheet, stretchy panels, fastening and landing strips, and airlaid-pulp absorbent core materials. The coverstock is either a longitudinally oriented polyester web treated with a hydrophilic finish, or a spun-laid polypropylene nonwoven used to contain and conceal an underlying core material. Examples include facing materials that cover the absorbent cores of diapers, sanitary napkins, and adult incontinence products. Coverstocks such as carded thermo-bonded nonwoven fabrics made with hydrophilic polypropylene fibers in thicknesses ranging from 16 to 30 g/m^2 are able to guarantee rapid fluid passage while maintaining contact with the surface of the skin and keeping it dry and comfortable.

Adult incontinence occurs in both men and women, and can range from light leakage of urine to a complete loss of control of urine and feces. The majority of adults with incontinence are women, and most of these experience light urinary incontinence. Consumer research has shown that many women use panty liners or sanitary napkins designed for menstruation to control small amounts of leakage because they are comfortable and discreet. Women with heavier leakage may use full-size sanitary napkins or pads specifically designed for female incontinence to achieve greater absorbency. Key consumer needs in this category include the need for protection and absorption, dryness, odor control, no leakage, etc. Other requirements include comfort and discretion, a sense of being secure and in control, and a need to forget about bladder weakness. The key product requirement is the ability quickly to acquire and store urine, which is more fluid, has a lower viscosity, and is expelled under more pressure than menstrual fluid.

Adults may lose the ability to control urine (bladder incontinence) or feces (fecal incontinence, also called bowel incontinence or anal incontinence) as a result of aging, injury, disease, or other factors that damage the nervous or muscular control needed to maintain continence. Common factors contributing to incontinence are pregnancy and childbirth, which can affect the strength of pelvic floor structures, and prostate surgery, which can damage nerves essential to the control of urination. Adults who are incontinent must struggle with both the physical aspects of dealing with the excreta and the social and emotional aspects relating to loss of dignity and fear that their condition may be betrayed to those around them by odor, leakage, or visibility of the incontinence product. In patients who are otherwise healthy and active, a discreet product allows the continuance of normal activities without fear of embarrassment. In patients with limited mobility or who are bedridden, the use of an absorbent product that protects the skin from excreta is particularly important, as these patients are at high risk of skin breakdown and bedsores.

Adult incontinence products take a number of forms, and the best product for a particular user will depend upon a number of factors, including sex, type and degree of incontinence, and lifestyle. All wearable disposable products are now made with fluff and SAPs, so that even small, thin products can contain large volumes of liquid and lock this inside so that skin remains dry.

6.9.4 Drapes and cover cloths

In the operating room, nonwoven composite materials are often used for surgical drapes and cover cloths, where the nonwoven fabrics are combined with polymeric films to produce a composite structure which is completely impermeable to bacteria while highly absorbent to both body perspiration and secretions from the wound. The addition of SAPs into the nonwoven fabric can significantly enhance the absorption capacities of these products.

6.9.5 Wipe cloths

Wipes are often made from tissue paper or spun-laced nonwoven fabrics of polypropylene in blend with viscose rayon in the range of 30–70 g/m^2, which may be soaked with an antiseptic finish. One of the most obvious requirements for a wipe cloth is its rapid absorbency, which may be achieved by the addition of a SAP component.

6.9.6 Contact lenses

Superabsorbent polymers were first used in the production of contact lenses in 1960, when a cross-linked hydrogel was obtained from 2-HEMA, which had a water absorption capacity of 80 g/g. This polymer has been applied in the production of contact lenses up to the present day. Owing to the presence of –OH groups, it is possible for oxygen to reach eyeballs through the gelled aqueous phase without any obstacles. When dry this material has the properties of hard organic glass, whereas after hydration it becomes soft and flexible, which is ideally suited for the production of contact lenses.

6.9.7 Drug administration

The gelled structures of SAPs can offer uninterrupted and controlled release of incorporated drugs into body fluids once a device is implanted, injected, or otherwise introduced into the body. The drugs can be released by two mechanisms. First, the release of small molecules takes place as a result of hydrogel swelling. Second, the drug-containing polymer matrix undergoes erosion, and diffusion of the drug into the surrounding environment is a function of the rate of biological decomposition.

6.9.8 Tissue engineering

Hydrogels made of SAPs are similar to soft tissues, and, due to their good biocompatibility and good permeability to oxygen and other water-soluble metabolites, they can create an efficient environment to ensure fast cell growth, which make them suitable for the preparation of matrices in tissue regeneration. It has been found that polyethylene oxide and polypropylene oxide copolymers are ideal substrates for the proliferation of cartilage cells.

6.9.9 Wound dressings

Hydrogel dressings made from SAPs are often used for burn wounds, and also for chronic wounds such as ulcers and bedsores. Owing to their high flexibility, they can be applied on wounds in any part of the body, and once applied it is also easy to remove the hydrogel dressings without causing pain to the patient. In addition to easy application and removal, it is recognized that skin regeneration and reconstruction can be accelerated by the application of hydrogel wound dressings, since the large amount of water in the hydrogel structure can assist in the creation of a moist environment on the wound, which is considered as ideal for cell proliferation and skin regeneration.

6.10 Hygiene implications in the use of absorbent hygiene products

Disposable superabsorbent hygiene products promote cleanliness by containing bodily fluids and excreta and allowing them to be disposed of in an appropriate way. Feces contain up to 10^{10} bacteria per gram of weight, and the fecal microorganism in infants is estimated to consist of 20 genera and 100 species. In addition to providing skin protection benefits, disposable superabsorbent diaper products provide better containment of excreta and reduce the environmental spread of pathogenic organisms. It has been found that significantly fewer objects were contaminated with fecal coliform bacteria in rooms in which children wore disposable superabsorbent diapers compared with double cloth diapers when no overclothes were worn.

The risk of diarrheal illness in daycare centers has been clearly linked with contamination of hands and items such as toys, so prevention of such contamination is an important step in controlling the spread of illness in this setting. It should be noted here that, when thrown away in the rubbish, a soiled disposable diaper is typically wrapped tightly using the diaper tapes, thus providing additional containment of the feces until safely disposed off by incineration or in a landfill. In contrast, reusable cloth diapers must be retained and laundered, allowing many opportunities for contamination of hands, surfaces, or other aspects of the environment with fecal organisms.

Hygiene issues for adult incontinence products are similar to those for baby diapers. In institutional settings, the ability to confine and dispose of urine and feces with a minimum of effort and without contamination of the environment is paramount. The use of modern disposable absorbent incontinence products should reduce the amount of time required to clean and care for incontinent elderly or frail patients, the costs and time needed to wash soiled clothing and bedding, and the risk of spreading illness in the institution through contamination. In active adults, the use of highly absorbent incontinence products allows continuation of a normal routine outside the home, without fear of leakage or embarrassment in public.

6.11 Summary

Superabsorbent polymers are widely used in the production of medical textile products with high absorption capacities, and are used in baby diapers, feminine hygiene products, adult incontinence products, and many other applications. In the medical textile industry, the novel water-holding structures of SAPs can be used in powder, fiber, and fabric forms in combination with other convention textile structures to develop high-absorbing products for the medical and healthcare industry, with many diversified applications such as functional wound dressings, drapes and cover cloths, drug administration, tissue engineering, etc.

Sources for further information

[1] Bartels VT, editor. Handbook of medical textiles. Cambridge: Woodhead Publishing Ltd; 2011.

[2] Zohuriaan-Mehr MJ, Kabiri K. Superabsorbent polymer materials: a review. Iran Polym J 2008;17(6):451–77.

[3] Alvarez O, Rozint J, Wisemann D. Moist environment for healing: matching the dressing to the wound. Wounds 1989;1:35–51.

[4] Buchholz FL, Graham AT. Modern superabsorbent polymer technology. New York: John Wiley & Sons; 1998.

[5] Buchholz FL, Peppas NA. Superabsorbent polymers: science and technology. Washington: American Chemical Society; 1994.

[6] Kiakamjornwong S, Punchareon P. Influence of reaction parameters on water absorption of neutralized poly-(acrylic acid-co-acrylamide) synthesized by inverse suspension polymerization. J Appl Polym Sci 1999;72(10):1349–66.

[7] Mitchell RL, Muller TE, Stevens HD, et al. Process for forming hydroxyethyl cellulose fibers having high water absorption and high water retention properties. US Patent 3670069; 1972.

[8] Askari F, Nafisi S, Omidian H, et al. Synthesis and characterization of acrylic-based superabsorbents. J Appl Polym Sci 1993;50:1851–5.

[9] Patel VR, Amiji MM. Preparation and characterization of freeze-dried chitosan-poly (ethylene oxide) hydrogels for site specific antibiotic delivery in the stomach. Pharm Res 1996;13:588–93.

[10] Kabiri K, Omidian H, Zohuriaan-Mehr MJ. Novel approach to highly porous superabsorbent hydrogels: synergistic effect of porogens on porosity and swelling rate. Polym Int 2003;52:1158–64.

[11] Omidian H, Hashemi SA, Sammes PG, et al. Modified acrylic-based superabsorbent polymers: effect of temperature and initiator concentration. Polymer 1998;39(15):3459–66.

[12] Desai AA. Special textiles used for manufacturing healthcare and hygiene products. Text Mag 2003;3:73–6.

[13] Williams M. From diapers to industrial textiles-fiber finishes that cover the surface. Int Fiber J 2001;6:74–5.

[14] Anand SC, Kennedy JF, Miraftab M, et al. Medical textiles and biomaterials for healthcare. Cambridge: Woodhead Publishing Ltd; 2005.

[15] Mathews A, Hardingham M. Medical and hygiene textile production. Rugby: Practical Action Publishing; 2011.

[16] McCarthy BJ. Textiles for hygiene and infection control. Cambridge: Woodhead Publishing Ltd; 2011.

[17] Zhong W. An introduction to healthcare and medical textiles. Paddyfield: DEStech Publications, Inc.; 2013.

Functional wound dressings

7

7.1 Introduction

Wounds are defined as skin defects caused by chemical, mechanical, electrical, or thermal injuries, or by the presence of an underlying medical or physiological disorder. Wound dressings are materials used to cover the wounds. Many types of wounds occur in everyday life, such as mechanical injuries like abrasions, lacerations, acute bullet wounds, knife cuts, bites, and surgical wounds, and various types of burns caused by thermal, chemical, electrical, or radiational injuries. Other types of wounds such as chronic ulcerative wounds, pressure sores, and leg ulcers occur more commonly among elderly people.

Wound dressings have been in use for as long as conventional textile materials. Their primary function is to avoid strike-through and protect the wounded site from contamination and further injuries. Wound dressings may not help accelerate the healing process, but they must not in any way delay the wound repair process through any mechanism, such as adhering to the wound bed, leaching toxic components, and causing wound maceration. They need to be easy to apply and easy to remove. These traditional requirements for wound dressings were largely met by the various types of gauzes and derivatives of gauze dressings, such as bandages, wax-impregnated gauze, woven, nonwoven, and knitted gauze, etc. Nonwoven materials laminated with perforated plastic film have been used to reduce both dressing adherence and the amount of fibrous residue, thus overcoming one of the main shortcomings of traditional wound dressings.

In the 1960s the science of wound dressings saw a breakthrough, when George D. Winter, while working at the Department of Biomechanics and Surgical Materials at the University of London, reported the results of a study on the treatment of pig wounds in an occlusive condition. It was found that when the wound was kept in a moist condition, as would result from the application of an occlusive dressing, epithelialization of the wound surface occurred much faster than if the wound was kept in an otherwise dry condition, which was then regarded as desirable. Further studies on humans confirmed that wounds heal much faster when they are kept in a moist condition. These early studies provided the scientific and medical foundation for modern wound-management materials, which are designed to provide a moist healing environment for the wound. The so-called "moist-healing" dressings were developed to create this moist interface between the wound and the dressing.

It is interesting to note that much of the research that led to the development of modern wound-management products was concentrated on the "moist-healing" principle, and in the 1980 and 1990s many moist-healing products, such as hydrocolloids, alginates, polyurethane foams, and hydrogels, were developed and launched in the

Medical Textile Materials. http://dx.doi.org/10.1016/B978-0-08-100618-4.00007-8

European and North American healthcare markets. The advent of an aging population together with the increased need for managing chronic ulcerative wounds provided a growing market for companies supplying wound-management products, and at the same time stimulated research and development efforts for smart wound-care materials that are more effective and more functional than traditional materials. There was a clear surge in the number of new products from 1996 onwards, as many new types of alginates, hydrogels, and other smart wound-care materials became commercialized.

7.2 The process of wound healing

The process of wound healing is a survival mechanism which commences immediately after the body sustains injury. It is the process by which tissues damaged or destroyed by injury are removed and the breach in tissue or skin integrity is restored. Wound repair is performed either by spontaneous regeneration of dermal tissue or by connective tissue repair. Both these occur regardless of how the injury has been sustained, or whether the injury is minor, major, confined to soft tissue trauma, or is a surgical wound. The mode of repair is primarily decided by the depth of the tissue involved. In minor injuries involving the epidermis, deep tissue regeneration is prompt, with reepithelialization taking place within 7–10 days. However, in wounds involving the loss of dermal tissue and skin adnexa such as hair follicles, sweat, and sebaceous glands, reepithelialization and regeneration cannot occur promptly, so the body relies upon the formation of connective tissue to fill the wound deficit.

Wound healing by primary intention involves the close approximation of the wound edges of a laceration or surgical wound by sutures, clips, or skin adhesive. Healing by secondary intention occurs when the wound edges cannot be approximated, in which case the wound is left open and the defect will slowly fill with connective tissue. Such wounds are slow to heal and prone to complications such as infection. Commonly, this type of wound healing occurs in patients with underlying comorbidities such as diabetic ulcers or pressure ulcers, and patients with postsurgical wound dehiscence. Wound healing by tertiary intention, which is often called "delayed primary intention," involves the wound being left open until any infection or contamination with nonviable tissue is removed; the wound edges are then approximated and healing continues as by primary intention.

The wound-healing process sets in motion a cascade of physiological responses which are coordinated and ordered, leading eventually to wound repair. Wounds that follow this process are termed "acute wounds," whereas wounds which fail to heal due to interruption in the ordered process or interference by an underlying disease are termed "chronic or indolent wounds." Typically, wound healing is a physiological process comprising several distinct phases: hemostasis, inflammation, proliferation of connective tissue, contraction, epithelialization, and remodeling.

7.2.1 Hemostasis

Within seconds of skin injury, vasoconstriction occurs to reduce blood loss. As blood spills into the wounded area, platelets come into contact with and adhere to the wall of the injured blood vessels and the exposed collagen within the ECM. This triggers platelets to release cytokines, growth factors, and numerous pro-inflammatory mediators, resulting in platelet aggregation and triggering the intrinsic and extrinsic coagulation pathways. This leads to fibrin clot formation, which seals the disrupted blood vessels, preventing further blood loss. The clot is sustained by the rapid intravascular accumulation of microvesicular tissue factor as microvesicles from white cells are attracted to the platelet surface. Growth factors produced by the platelets initiate the healing cascade. Platelet-derived growth factor (PDGF) initiates the chemotaxis of neutrophils and macrophages, smooth muscle cells, and fibroblasts, while also stimulating mitogenesis of smooth muscle cells and fibroblasts. Transforming growth factor-beta (TGF-b) attracts macrophages into the wound area and stimulates them to produce additional cytokines, including fibroblast growth factor, PDGF, tumor necrosis factor-alpha, and interleukin-1. In addition, TGF-b further enhances the fibroblast and smooth muscle chemotaxis and modulates the expression of collagen and collagenase. The purpose of PDGF and TGF-b is to stimulate a response from the matrix to produce cells in preparation for the rapid deposition of new connective tissue to rebuild the extracellular matrix (ECM). The clinical picture of hemostasis is clearly demonstrated by wounds such as skin graft donor sites which heal by regeneration.

7.2.2 Inflammation

Tissue injury and the activation of the coagulation cascade stimulate the release of vasoactive cytokines such as prostaglandins, histamine, and other amines from the granules released by mast cells as a response to injury, which results in increased local vasodilation and capillary permeability. The vessels become more permeable, allowing the migration of monocytes into the wound bed. Serous fluid also leaks into the wound bed and surrounding tissue, creating edema. These are some of the characteristic signs of inflammation. The patient is likely to complain of increasing pain 24h after wounding or surgery, hence the need for good analgesic provision. During the period of inflammation neutrophils accumulate within the wound area—their primary role being to act as a "wound cleanser." They are initially attracted to the wound area by various chemotactic agents released from bacteria. Once in the wound, they ingest the bacteria via phagocytosis. Neutrophils filled with bacteria and debris produce the appearance of pus or slough in a wound bed.

It should be noted that the normal inflammatory phase of wound healing should not be confused with the abnormal inflammation secondary to clinical infection, which tends to produce a raised temperature and raised white cell count.

7.2.3 Proliferation of connective tissue

There is a wide overlap between the phase of inflammation and the next phase of proliferation, when it is possible to see within a wound bed both slough and necrotic tissue and burgeoning granulation tissue. During the proliferative phase the wound defect is filled with highly vascular connective tissue, commonly referred to as "granulation tissue." This process is instigated at the point of wounding, when as a result of loss of vascularity in the wound bed, the wound environment has a low pH, reduced oxygen tension, and increased lactate. This initiates the release of vascular endothelial growth factor (VEGF), basic fibroblast factor, and TGF-b, all of which activate neovascularization or angiogenesis. The low oxygen tension caused by lack of blood supply actually initiates the release of hypoxia-inducible factor (HIF), which regulates the expression of VEGF. Endothelial cells in the venules adjacent to the wound bed initiate capillary production by projecting pseudopods into the wound bed. Simultaneously, the endothelial cells produce enzymes to break down the adjacent ECM to create tissue defects into which the capillary vessels infiltrate to form a network; this stimulates the formation of arterioles and thus vascularity is restored. As new blood vessels intrude into the wound environment, the oxygen tension rises and oxygen binds on to the HIF, blocking its activity, which in turn leads to decreased production of VEGF. The process of angiogenesis is interdependent on the production of a new ECM, which acts as a scaffold to support the newly formed blood vessels. The predominant cells responsible for the formation of the ECM are fibroblasts, which are attracted into the wound bed by cytokines produced by macrophages and then transformed into wound fibroblasts, which have decreased proliferative behavior but increased collagen production behavior.

Collagen synthesis to form the new ECM is a multilayered process. During the initial phase of wound healing, the fibrin clot supports the migration of cells into the wound. This fibrin matrix is gradually replaced by material comprised mainly of fibronectin and hyaluronic acid, both of which promote cell migration and proliferation. Fibroblasts attach to the provisional fibrin matrix and begin collagen production. The early so-called "type 3 collagen" comprises approximately 30% of granulation tissue and does not contribute to the tensile strength of the wound. Hence any wound healing by secondary intention requires supportive dressings during this stage—for example, an abdominal support belt for a patient with a dehisced abdominal wound. As collagen matures, type 3 collagen is converted to "type 1 collagen," normally found in dermal tissue.

In wound healing by secondary intention it takes a relatively long time to produce the volume of connective tissue required to fill the injured part, where the granulation tissue is formed by the combined process of neovascularization and collagen production. Granulation tissue is slow to form in patients with poor nutritional status, and appropriate wound management is vital for the healing process. Topical negative pressure can provide an occlusive environment while increasing the rate of neovascularization. Absorbent dressings such as nonwoven alginate or polyurethane foams are used to absorb excess exudate and prevent skin maceration.

7.2.4 Contraction

The phenomenon of contraction only occurs in open wounds, where it is desirable since less connective tissue is required to fill the wound deficit and healing will take place more quickly. There are three possible mechanisms for wound contraction. The first theory suggests that modified fibroblasts (myofibroblasts) are responsible for generating contractile forces which pull wound edges together. A second theory suggests that wound fibroblasts act collectively to contract the connective tissue matrix. A third theory proposes that the newly formed collagen fibers are compacted by cellular forces producing a pulling force on surrounding tissues. The degree of contraction within the wound bed is determined by the mobility of the surrounding tissue: in areas such as the abdomen the overlying tissue is very mobile and contracture is easily achieved, in contrast to skin over bony prominences. Wound contracture can be undesirable in some wounds when cosmetic deformities may arise. In such cases, wound healing by secondary intention is undesirable and the wound defect should be filled by a skin graft if it is superficial, and by a myocutaneous flap if it is deep.

7.2.5 Epithelialization

In this final stage of visible wound healing, epithelial cells migrate from the wound edges to resurface the wound defect. In wound healing by primary intention, epithelialization occurs concurrently with connective tissue deposition. In healing by secondary intention, this process is delayed until the wound defect is filled with granulation tissue. Epithelial cells can only migrate over a moist, vascular wound surface, and are inhibited by a dry or necrotic wound surface. Lateral migration proceeds until the defect is covered, before the cells resume upward migration and differentiation and the epidermis regains its normal thickness and stratification.

7.2.6 Remodeling

The remodeling phase may extend to a year or more, during which fibroblasts regulate the process of wound matrix breakdown by matrix metalloproteinases and synthesis of new ECM. This slow process increases the tensile strength of the wound, but scar tissue never has more than 80% of the tensile strength in unwounded tissue. Occasionally, in the process of remodeling, an imbalance may occur, during which the processes of matrix degradation and synthesis are disrupted, resulting in abnormal scar formation such as hypertrophic or keloid scarring.

In general, wound healing is a well-coordinated process provided that the individual patient is nutritionally robust, hemodynamically well, and biochemically stable.

7.3 Functional requirements for modern wound-care materials

Wounds differ from one type to another, and the functional requirements for wound dressings and wound-management materials can only be specified according to the specific types of wounds. In defining requirements of a wound dressing, it is also important to observe that for a specific type of wound the requirement for the dressing is different at different stages of the healing process.

According to their physiological conditions, wounds can be classified into five types, each with a symbolic color code: necrotic wounds, sloughy wounds, granulating wounds, epithelializing wounds, and infected wounds, which are characterized by their black, yellow, red, pink, and green appearances respectively. Figure 7.1 illustrates the five types of wounds.

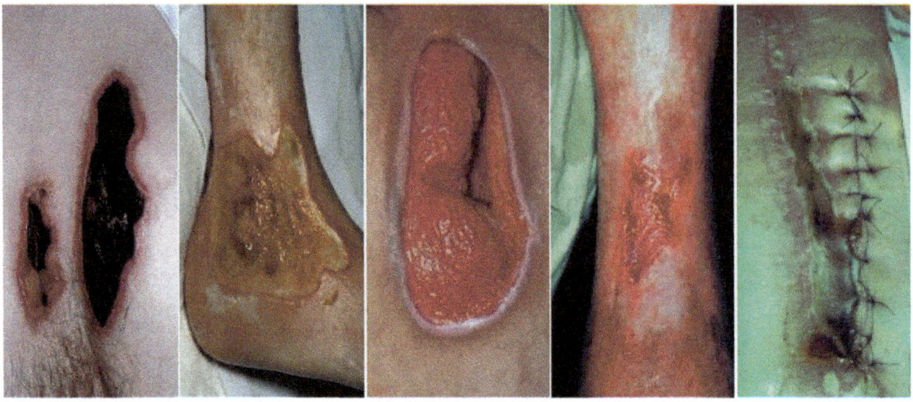

Figure 7.1 A schematic illustration of the five types of wounds: (a) necrotic; (b) sloughy; (c) granulating; (d) epithelializing; (e) infected.
Source: Smart Textiles for Medicine and Healthcare (Chapter 2: Smart woundcare materials). Woodhead Publishing, 2007.

Because of their differences in the level of exudate, level of microbial contamination, and other physiological characteristics, the management procedure differs considerably for the five different types of wounds. For necrotic wounds, one of the main aims of applying a dressing is to facilitate the separation of the dead tissue from the underlining healthy tissue so as to enable a normal wound-healing process. If exposed to a relatively dry atmosphere, as found in a hospital ward or a centrally heated room, dead tissue rapidly loses moisture and becomes dehydrated. As it does so, it shrinks and progressively becomes olive-green or black in appearance. It is also hard and dry to touch. In this condition, autolysis is inhibited and separation of the necrotic tissue may be delayed indefinitely. For the treatment of necrotic wounds, it is essential that the dressings must be able to prevent the dehydration process—indeed, they should be able to facilitate a hydration process for the dead tissue so that the autolytic debridement process in which dead tissue is separated from healthy tissue can take place. Necrotic wounds are ideally treated with hydrogels and/or hydrocolloid dressings.

For yellow, sloughy wounds, such as burns, leg ulcers, and pressure sores, where a necrotic covering has been removed, a glutinous yellow covering normally develops on the wound surface. This is not dead tissue, but a complex mixture of fibrin, protein, serous exudate, leucocytes, and bacteria. To treat this type of wound, the sloughy mess must first be properly cleaned or debrided. If the slough is moist, an alginate dressing can be used to absorb the exudate; if the slough is dry, hydrocolloid or hydrogel dressings may be used.

Normal wound healing progresses from necrotic and sloughy phases to the granulating phase, when the wound is covered by granulation tissue composed of collagen and proteoglycan in a complex of protein and polysaccharides with salts and other colloidal materials. These produce a gel-like matrix which is contained within the fibrous collagen network, with a highly vascular nature that gives it a characteristic deep pink color. Granulating wounds vary considerably in size, shape, and the amount of exudate they produce. As a result, no single dressing will be suitable for use in all situations. Depending on the particular shape and the amount of exudate, alginate, hydrocolloids, or hydrogels may be used.

The final phase of the wound-healing process is the epithelialization of the wound surface. With a few exceptions, superficial or epithelializing wounds tend not to produce large quantities of exudate. Traditionally, these wounds have been dressed with paraffin gauze and cotton tissue, but sometimes alginate and hydrocolloids can be used. It is important to recognize that at this stage the tissue is soft and fragile, and therefore any dressings used on these wounds must not adversely disturb the delicate tissue structure. In particular, dressings must not unnecessarily adhere to the wound. For epithelializing wounds, alginate, hydrocolloid, vapor-permeable film, silicon-coated film, and knitted viscose gauze may be used as the primary dressings.

Infected wounds generate a large amount of exudate and also unpleasant odor. The treatment of infected wounds therefore comprises absorbing exudate, containing odor, and controlling microbial contamination. Dressings that in part contain activated charcoal can be used to absorb odor, while the odor-generating bacteria can be controlled by using dressings with an antimicrobial function, such as those with chlorhexidine, silver compounds, or iodine.

In summary, the main functions of wound dressings are as follows.

1. **Physical barrier**
   One of the principal functions of wound dressings is to avoid strike-through. In addition to their aesthetic purposes of hiding the wound, dressings also help to separate the wound surface from the atmosphere and prevent the wound from bacterial contamination and further physical damage to the tissue.
2. Fluid control
   The ability to absorb fluid from a highly exuding wound or to donate moisture to a dry wound is one of the main functions of a dressing.
3. Microbial control
   For infected wounds, it is important that bacteria are contained by appropriate methods.
4. Hemostatic effect
   For acute surgical wounds and traumatic wounds, it is important that bleeding is stopped as early as possible to prevent blood loss. Appropriate wound dressings can help blood clotting.
5. Odor management
   Wounds quite often produce an unpleasant and obnoxious odor. When this occurs, dressings must be able to contain the odor.
6. Space filler
   For deep cavity wounds, it is important that the wound is kept open by filler materials so that the healing process can take place from the bottom upwards and unnecessary closure of the wound before the whole cavity is healed can be prevented.
7. Debridement
   The removal of dead necrotic tissue is essential in facilitating the normal wound-healing process. Wound dressings can accelerate the debridement process by providing the appropriate moisture, pH, temperature, and other conditions that are ideal for the autolytic debridement process.

8. Low adherence
   Adherence of the whole or part of the dressing to the wound surface is a major problem in wound management, often causing trauma on removal of the dressing. A low-adherent dressing can help lower or eliminate adherence to wound.
9. Scar reduction
   For large wounds, scar formation presents a major aesthetic problem for the patient. Any dressing that can reduce or prevent scar formation can give a great benefit to the patient.
10. Metal ion metabolism
   A number of metal ions, such as iron, zinc, copper, magnesium, and selenium, play important roles in cellular activities. Deficiency in any metal ion delays wound healing. Apart from systemic intake of these metal ions in the diet, they can be administered through the topical use of appropriate wound dressings.
11. Pressure regulation
   The arterial and venous blood flow requires an appropriate pressure, and a pressure sore develops when body parts are subjected to sustained high pressure. Elastic bandages can help regulate body pressure and assist the wound-healing process.
12. Wound-healing acceleration
   Wound healing is a complex physiological process, and dressings often play a minor role in the overall rate of healing. However, when combined with a number of factors, the appropriate use of dressings can accelerate the wound-healing process.

7.4 Functional wound dressings

Because of the similarity in applications and functional requirements, wound dressings are often made from materials similar to those used in traditional textile fabrics. Indeed, traditional gauze materials are made from cotton yarn, using a simple woven technique. For different applications, traditional wound dressings use many different materials and have different types of textile structures. For example, the British Pharmacopoeia classifies surgical dressings into four types.

1. X-ray-detectable absorbents.
2. Extensible bandages.
3. Tubular bandages.
4. Wound dressings and medicated bandages.

It is often difficult to draw a line between the so-called traditional dressings and modern functional dressings, although the development of the moist-healing concept has often been credited as the starting point for modern wound dressings. Since the first generation of moist-healing dressings were developed in the late 1970s, more and more advanced dressings have been marketed by a number of wound-care companies in Western Europe and North America, expanding gradually to the rest of the world. As an illustration of the diversified modern wound-care materials, the British Drug Tariff lists the following 10 categories of dressings.

1. Low-adherence dressings.
2. Gauze-impregnated dressings.
3. Dextranomer paste pad and dressing.

4. Alginate dressings.
5. Hydrocolloid dressings.
6. Hydrogels.
7. Vapor-permeable adhesive film dressings.
8. Polyurethane foam dressings.
9. Zinc paste bandages.
10. Iodine-containing dressings.

These products are often used alone or in combination on wounds with diversified physiological backgrounds and at different stages of the healing process. Many types of functional materials have been developed, and some of these are introduced below.

7.4.1 Alginate fibers and wound dressings

As mentioned in Chapter 3, alginate is a natural polysaccharide extracted from brown seaweeds. Alginate fibers can be made by extruding aqueous sodium alginate solution into an aqueous calcium chloride coagulation bath, using a simple wet-spinning process. On contact with wound exudate, alginate fibers have a unique ion-exchange property whereby the calcium ions in the fiber exchange with the sodium ions in the body fluid, and as a result part of the fiber becomes sodium alginate. Since sodium alginate is water soluble, this ion exchange leads to swelling of the fiber and in-situ formation of hydrogel on the wound surface. Figure 7.2 shows photomicrographs of alginate fibers when placed in contact with water and saline respectively. A highly

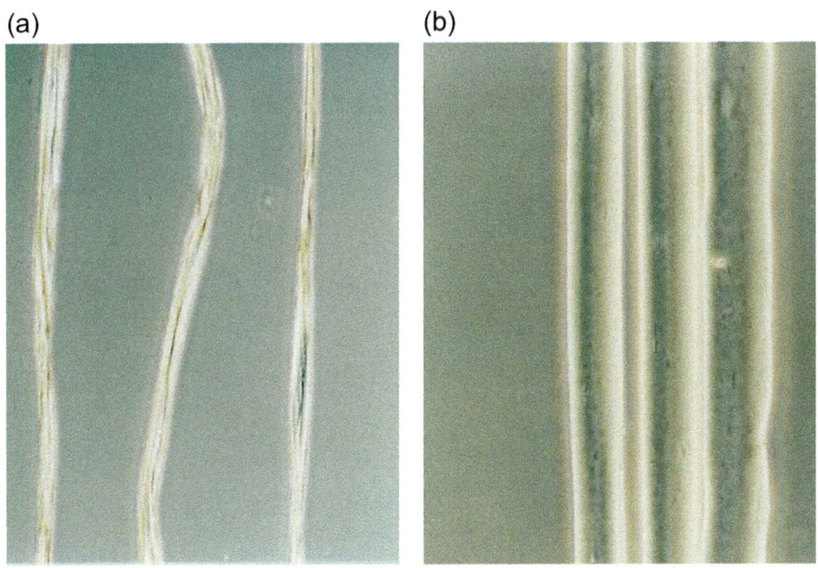

Figure 7.2 Photomicrographs of alginate fibers: (a) wet in water; (b) wet in saline.
Source: Smart Textiles for Medicine and Healthcare (Chapter 2: Smart woundcare materials). Woodhead Publishing, 2007.

swollen structure can be obtained by placing calcium alginate fibers in contact with aqueous solutions containing sodium ions.

The gelling properties of alginate fibers are particularly important in the production of functional wound dressings. First, as the fibers hold more water in a swollen structure, the dressings are capable of absorbing more wound exudate, hence extending the duration of dressing use. Second, when the fibers absorb water into their structure and swell, the spaces between the fibers in the dressing are closed, thus prohibiting liquid from lateral spreading and preventing maceration of the areas surrounding the wound surface. This unique feature is known as the "gel-blocking" property of alginate wound dressings. Due to the release of calcium ions during the ion-exchange process, alginate wound dressings are also known to have excellent hemostatic properties.

Figure 7.3 shows the chemical structures of β-D-mannuronic acid (M) and α-L-guluronic acid (G), which form the linear polymeric chains. These two monomers are stereochemically very different, as a result of their difference at C-5. As shown in Table 7.1, alginates extracted from different species of seaweed differ in M and

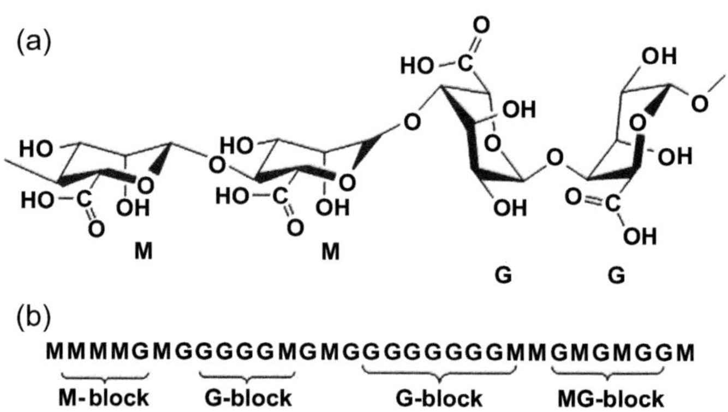

Figure 7.3 The chemical structures of β-D-mannuronic acid (M) and α-L-guluronic acid (G) and their distribution in the polymeric chain.

Table 7.1 The mannuronic acid and guluronic acid contents of alginate extracted from different types of brown seaweeds

Type of seaweeds	Mannuronic acid (%)	Guluronic acid (%)
Ascophyllum nodosum	60.0	40.0
Laminaria digitata	59.0	41.0
Laminaria japonica	69.3	30.7
Macrocystis pyrifera	61.0	39.0
Laminaria hyperborean, fronds	56.0	44.0
Laminaria hyperborean, stems	30.0	70.0
Lessonia nigrescens	62	38

G contents, resulting in the differences in the physical properties of different types of alginate fibers and wound dressings. During clinical applications, alginate wound dressings with a high M content are known to have good gelling properties, while those with a high G content possess good wet integrity.

7.4.2 Chitosan fibers and wound dressings

Chitosan is the deacetylated derivative of chitin, which is the second most abundant natural polymer and exists widely in cell walls of fungi and crustacean shells. Chitin is known to have accelerating effects on the wound-healing process, and fibers, mats, sponges, and membranes have been made from chitin for use as wound-dressing materials.

By deacetylating chitin in an alkali solution, the resultant chitosan has primary amine groups in its structure, and it is easy to prepare a spinning solution by dissolving chitosan in aqueous acidic solutions. Chitosan fibers are made by extruding the spinning solution through fine holes into an alkaline coagulation bath, in which chitosan precipitates out in the form of long filaments. After washing, stretching, and drying, these fibers can be used for the production of wound dressings.

Nonwoven dressings made from chitosan fibers are now widely used for managing burns and other types of wounds. Chitosan-based wound dressings are biocompatible, biodegradable, and nontoxic. They are known to have natural antimicrobial properties and are able to promote wound healing through the release of glucosamine, a by-product of the enzymatic degradation process.

To improve their absorption properties, chitosan fibers and wound dressings can be treated with chloroacetic acid to convert them into partially carboxymethylated chitosan. As can be seen in Table 7.2, fibers with various degrees of carboxymethylation can be prepared by controlling the ratio between the weight of fiber and the chloroacetic acid. When the degree of carboxymethylation is below 50%, the carboxymethylated chitosan fibers can absorb a large amount of liquid while maintaining fibrous structure, making them highly valuable in the production of moist-healing wound dressings.

Table 7.2 **Absorption properties of carboxymethylated chitosan fibers**

Degree of carboxymethylation (%)	Absorbency (g/g)	Gel swelling ratio (g/g)
0	10.8	1.20
4.33	11.4	2.76
20.93	15.0	6.65
41.72	22.1	12.47
62.22	6.10	3.41

7.4.3 Superabsorbent cellulosic fibers and wound dressings

Cellulosic fibers such as cotton and viscose rayon can be treated with chloroacetic acid to make highly absorbent carboxymethyl cellulose fibers. When partially carboxymethylated, the carboxylic acid groups in the fibers are able to absorb a large amount of water into the fiber structure, hence the treated fibers are capable of a high degree of swelling when wet in water.

When the carboxymethylated cellulose fibers are made into nonwoven dressings, the resultant product possesses both the soft and fine features of a nonwoven fabric and also the superabsorbency derived from the carboxymethylation treatment. In the presence of wound exudates, carboxymethylated cellulosic dressings can form a cohesive gel sheet, which facilitates their use under compression bandages. The nonadherent features of these dressings lead to significantly less pain on dressing change. In clinical applications these products have a range of benefits, such as superior exudate absorption and retention, improved handling characteristics, and enhanced vertical wicking, which minimizes the potential for maceration.

7.4.4 Oxidized cellulose fibers and hemostatic gauzes

Oxidized cellulose can be obtained by partial oxidation of the primary hydroxyl groups on the anhydroglucose rings of cellulose. Oxidized cellulose was studied for the first time in the late 1930s, with nitrogen dioxide as the oxidant for the selective oxidation of the primary hydroxyl groups in cellulose. Since then, both gaseous and liquid nitrogen dioxide oxidation processes have been commercialized.

Within the range of 16–24% carboxylic acid content, oxidized cellulose has a pH of approximately 3.1. The material is biocompatible, bioabsorbable, and hemostatic. Based on these unique properties, Johnson & Johnson pioneered an industrial-scale oxidation process by using nitrogen dioxide to convert regenerated cellulosic fibers into oxidized regenerated cellulose (ORC) fibers. When knitted into a flexible fabric, ORC can be used as an absorbable hemostat in various surgical operations where blood loss should be controlled.

ORC can be prepared in the following process. First, NO_2 is dissolved in CCl_4 to prepare a 20% (wt) NO_2/CCl_4 oxidant solution. Then knitted regenerated cellulose fabric is added to a round-bottomed flask containing this oxidant, with a proportion between knitted fabric and oxidant of 1:42.6 (g/mL). The mixture is stirred constantly and the reaction temperature is kept at 19.5 °C for 40 h. After the reaction, the knitted fabric is washed three times with CCl_4 before being washed three times with an aqueous solution containing 50% (v/v) ethanol. Finally, the product is washed three times with pure ethanol and freeze dried at −50 °C in vacuum for 48 h.

7.4.5 Polyurethane film and foam

Semipermeable polymeric coatings have been used in the textile industry for making breathable clothing. In the wound-management industry, semipermeable films are also used to make dressings that can allow gaseous exchange between the wound and the surrounding environment while preventing air-borne bacteria from contaminating the wound. The semipermeable nature of the film allows a high rate of moisture vapor transmission through the film, thereby reducing the build-up of wound exudate under the film.

Modern semipermeable film dressings are usually made of polyurethane of various compositions, and the moisture vapor transmission rate (MVTR) can be modified by modifying the polymer and the film structure. An MVTR of 3000 $g/m^2/24h$ or greater can be achieved with the latest high-tech films. These film dressings can be used in the treatment of superficial partial-thickness burns and early decubitus ulcers. They are light and resilient, which is ideal for the prevention of skin damage by friction. Because of their protective properties, these products are also widely used as secondary dressings in conjunction with hydrogels, hydrocolloids, and alginates.

Polyurethane foams are soft and porous materials that can be used for wound-contacting primary dressings or as secondary dressings, utilizing their relatively high strength and flexibility. The polyurethane can be made from hydrophobic or hydrophilic monomers, resulting in foams with varying characteristics of porosity and fluid-handling capability. When placed on wet wounds, fluid is absorbed into the foam by capillary action and transferred across the dressing. When placed on relatively dry wounds, the polyurethane backing layer reduces moisture vapor loss and helps to prevent dehydration of the wound surface.

7.4.6 Hydrogels

Hydrogels are cross-linked polymeric networks swollen in biological fluid. They are widely used in drug delivery and tissue/organ repairs. Two types of hydrogels are usually available: sheet and amorphous.

With sheet hydrogel, the hydrophilic polymers, typically polyacrylamide or polyethylene oxide, are partially cross-linked to form a membrane with sufficient water-holding hydrophilic sites. Typically a hydrogel contains about 96% water. When applied to wound sites such as dermabrasions, minor burns, and skin donor sites, they relieve pain and reduce trauma on both application and removal of the dressings. Hydrogels can be dried to form a dehydrated hydrogel, which has a higher absorption capacity than the hydrated form.

Amorphous hydrogels can be made from a number of water-soluble polymers, such as cross-linked carboxymethyl cellulose, modified starches, alginate, and pectin. Unlike sheet hydrogel, which has a three-dimensional structure, these gels are thick, viscous fluids. When absorbing wound exudate, they swell until they lose all cohesive properties. Amorphous hydrogels are excellent donors for water. They are ideally used for dry and sloughy wounds whereby their water-donating properties assist the autolysis process for debriding and cleansing slough and black necrotic tissue from the underlying healthy tissue. Amorphous hydrogels are also widely used in the treatment of cavity wounds, where their easy flowing properties can be used to fill a deep cavity to avoid wound closure from side to side.

7.4.7 Hydrocolloids

Hydrocolloid wound dressings are among the first type of high-tech modern dressings. They are typically made of hydrophilic polymeric granules dispersed in an elastic adhesive matrix. Typically, the granules are hydrophilic polymers such as sodium carboxymethyl cellulose, pectin, gelatin, and sodium alginate, while the adhesive matrix is typically polyisobutylene. The hydrocolloid and the adhesive form a homogeneous mixture such that the granules are uniformly dispersed in the adhesive matrix. The final dressing is usually composed of the hydrocolloid matrix cast on a sheet of polymeric membrane or film, which can be occlusive or semipermeable for different applications. On contact with wound exudate, the hydrophilic granules absorb the fluid to form a hydrogel, while the adhesive material provides a tack that keeps the dressing adhered to the wound. Since moisture transportation is relatively slow through the hydrocolloid matrix, the wound surface is covered by a moist contacting layer which assists in healing. The dressing is easy to remove, and provides good protection against bacterial contamination of the wound.

7.4.8 Activated carbon

The production of wound odor can represent a major problem for patients and their carers. Wounds most commonly associated with odor production include leg ulcers and fungating lesions of all types. The smell from these wounds is caused by a cocktail of volatile agents that includes short-chain organic acids, such as n-butyric, n-valeric, n-caproic, and n-caprylic acids produced by anaerobic bacteria, together with a mixture of amines and diamines such as cadaverine and putrescine that are produced by the metabolic processes of other proteolytic bacteria.

The most effective way of dealing with malodorous wounds is to prevent or eradicate the infection responsible for the odor, for example through the administration of systemic antibiotics or topical antimicrobial agents. If the formation of odor cannot be prevented, it may be necessary to use a dressing that can absorb the smell. Dressings containing activated carbon can be used for the treatment of these malodorous wounds.

Activated carbon cloth can be produced by carbonizing a suitable cellulose fabric by heating it under carefully controlled conditions. During this process, the surface of the fibers breaks down to form small pores. These greatly increase the effective surface area of the fibers and hence their ability to remove unpleasant smells, since the molecules that are responsible for the production of odor are attracted to the surface of the carbon fibers and held there by electrical forces.

7.4.9 Low-adherent dressings

One of the major problems with traditional gauzes is the difficulty in removing them upon healing of the wound. Fibrous matters from the dressing tend to mix with exudate and blood. When the wound heals and dries, the fibers often get stuck to the dry escar. Bleeding of the new tissue is common in these cases, and the trauma in removing traditional gauze from large wounds often poses a serious problem. Low-adherent dressings were developed to prevent this problem by using a primary wound-contacting layer that separates the exuding wound from the absorbent dressing layer. A typical low-adherent dressing is the so-called "tulle gras" invented during the First World War. These dressings are usually made of cotton gauzes impregnated with a coating of paraffin, which provides low adherence to the wound surface. Paraffin gauzes are frequently medicated to enhance their performance.

For heavily exuding wounds, the poor absorption properties of tulle gras tend to reduce the fluid uptake, and skin maceration occurs as a result of the build-up of fluid under the dressing. Other forms of low-adherent dressings were developed to tackle this problem. Most of the modern low-adherent dressings are perforated polymeric films, with the pores acting as the channel for exudate to pass through to the absorbent layer. Knitted gauzes made of continuous viscose rayon filaments are also used as a low-adherent dressing for the treatment of such heavy-exuding wounds as leg ulcers. However, since it is easy for new tissue to grow into the pores of the perforated film or knitted structure, the removal of these dressings may still cause tissue damage. In this respect, modern dressings such as hydrogels, hydrocolloids, and alginate offer far superior properties for both good fluid handling and low adherence.

7.4.10 Interactive wound dressings

Interactive wound dressings can interact with cells or matrix proteins in the wound bed to promote healing. For example, alginate wound dressings can exert a bioactive effect by activating macrophages within the chronic wound bed to generate pro-inflammatory signals, such as tumor necrosis factor-alpha and interleukin-1, -6, and -12. This may then initiate an inflammatory response characteristic of healing wounds.

Another example of an interactive dressing is Promogran, which is a sterile, freeze-dried matrix made up of collagen and ORC. In the presence of wound exudate, the matrix absorbs liquid and forms a soft, conformable, and totally biodegradable gel. This gel binds and inactivates matrix metalloproteinases, which when present in excessive levels have a detrimental effect on wound healing as they damage regenerating tissue. The gel also binds growth factors secreted by macrophages and fibroblasts in the wound bed, protecting them from degradation by proteases. As the gel is digested during the course of healing, the growth factors are released back into the wound bed in their active form, thereby promoting the healing process.

7.4.11 Tissue-engineered "skin equivalents"

Surgical grafting of split-thickness autologous skin is the standard method for rapid closure of full-thickness burn wounds. Modern tissue engineering has now made possible grafts using either sheets of fibroblasts in a biodegradable matrix or cultured keratinocyte sheets. Superior results can be obtained when both dermal and epidermal components are combined in a bilayer skin equivalent. In addition, artificial skins can be developed by combining cells with either synthetic matrices, such as polyglycolic acid mesh, or natural biological substrates such as collagen and glycosaminoglycans.

7.4.12 Antimicrobial wound dressings

The spread of antibiotic-resistant strains of microorganisms such as methicillin-resistant *Staphylococcus aureus* represents an ever-increasing threat to the health of vulnerable people throughout the world. To overcome microbial infection, many dressings now exploit "bioactive" properties to promote healing and control infection by using sustained release of iodine, silver, and other antimicrobial compounds. More details on antimicrobial dressings are given in Chapter 10.

7.4.13 Composite wound-care products

After many years of research and development, especially since the widespread use of moist-healing products in the 1990s, modern wound dressings are now much more functional and smarter than traditional products, such as cotton gauzes and absorbent swabs. Compared to traditional products, the new generation of functional wound-care products tends to be easy to use and cost effective. In achieving these functionalities, modern wound-care products also adopt many novel composite structures. For example, in the management of leg ulcers a four-layer system has proven effective in applying compression as well as exudate management to chronic ulcers; this is achieved by first applying a wool bandage from the base of the toes to just below the knee joint, followed by the application of a crepe bandage and then an elastic compression bandage, and finally wrapping in a cohesive layer.

Many other types of composite wound-care products have also emerged. In general, these products are composed of three key components: the wound contact layer, the functional layer, and the retention layer. Figure 7.4 shows a schematic illustration of modern composite wound-care products.

In a three-layered structure, the contact layer serves to provide a low- or nonadherent interface between the wound surface and the dressing. This should allow wound exudate to pass into the functional layer while preventing the adherence of the dressing to a drying surface at the end of the healing process. At the same time, the contact layer also stops the release of loose fibers or particles from the dressing into the wound site. Materials that can be used as a wound contact layer include polyamide nonwoven, polyurethane foam, perforated plastic film, knitted viscose filament yarn, silicone gel, etc.

The functional layers vary greatly for different products because of the variations in their intended uses. For the majority of wounds, the main issues in wound management are to contain exudate, control microbial growth, and combat odor. In these cases, a layer of superabsorbent material, antimicrobial substance, and activated carbon fabric respectively can be laminated into the wound dressing to provide the functional layer.

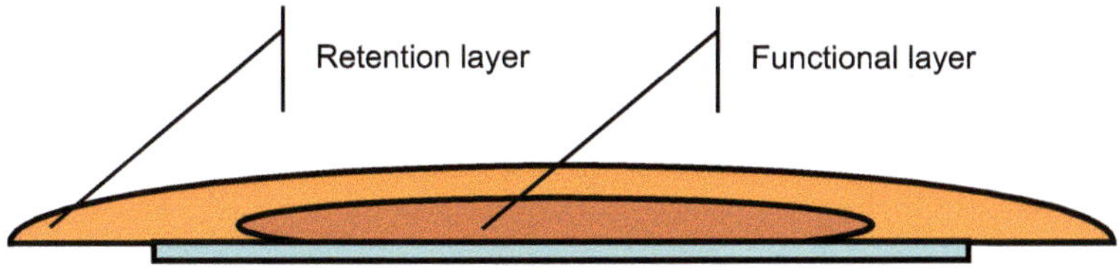

Figure 7.4 A schematic illustration of modern composite wound-care products.
Source: Smart Textiles for Medicine and Healthcare (Chapter 2: Smart woundcare materials). Woodhead Publishing, 2007.

The retention layer is used for two main purposes: to secure the dressing on to the wound edge and to provide physical protection of the wound surface. Polyurethane films and hydroentangled nonwoven fabrics are often used as retention layers. These materials are soft and flexible, thus making it easy to apply the dressings on to curvy areas of the body, such as around the shoulders and elbows. In addition, these materials are breathable, allowing oxygen to penetrate into the wound and moisture to evaporate from under the dressing, thereby extending the use duration of the product.

7.5 Cost versus effectiveness

Cost is an important issue in the development of functional wound-care materials, in addition to clinical efficacy and treatment effectiveness. In considering the cost of new treatment regimens, it is important to evaluate the cost not only in terms of direct treatment costs but also in terms of length of initial hospital stay, requirements for home care, additional bandaging regimens, and quality of the overall outcome. While advanced functional wound-care products are often more expensive than traditional products, in many cases this additional cost is justifiable. With respect to novel dressing types, it has been shown that not only are some of the new products cost effective, but they have also proven to be extremely beneficial in terms of their ability to reduce pain, odor, and leakage from wounds. For example, the cost effectiveness of tissue-engineered skin replacement has been borne out by evidence from a number of studies. When treating venous leg ulcers, the annual estimated medical cost of patient management with a tissue-engineered skin replacement equated to $20,041, compared to $27,493 for traditional therapy using Unna's boots. Despite the initial outlay being greater, the use of modern functional wound dressings reduced the overall treatment cost considerably. In addition, patients' quality of life can be improved.

7.6 Summary

Today, wound-care practitioners have at their disposal a wide range of functional wound dressings. These can be grouped under different product technologies, such as hydrocolloid dressings, alginate dressings, hydrogels, and foam dressings. Each of these product groups contains various brands that often have markedly different performance characteristics. As part of an overall wound-management plan, these functional dressings can help facilitate fast wound healing by providing the optimal environment for healing to proceed. They are also able to deal with odor, leakage, maceration, pain, infection, and other problems for wounded patients.

It should be pointed out that although wound dressings may not be able to promote the healing process significantly, the use of inappropriate dressings can often cause delayed wound healing. This may result in various complications of poor wound management.

1. Delayed healing and wound deterioration.
2. Increased risk of local or systemic infection.
3. Increased demand for nursing time and increased dressing costs.
4. Damage to the wound surface.
5. Damage to the surrounding skin.
6. Failure to control odor.
7. Detrimental effect on the quality of patients' lives.

In order to provide the best solutions for wound-care practitioners, new types of advanced wound dressings and wound-management materials are being developed to provide the means of ensuring the best possible patient outcome. The aims of the new developments can be summarized into the following three areas.

- Efficacy of the material.
- Effectiveness of the product.
- Efficiency of the treatment.

These three "E"s symbolize the goal-oriented approach in the current and future development of wound-management materials and products.

Sources for further information

[1] Sussman C, Bates-Jensen B. Wound care: a collaborative practice manual for health professionals. Oxford: Blackwell Publishing; 2012.
[2] Qin Y. Functional wound dressings. Beijing: China Textile Press; 2007.
[3] Winter GD. Formation of the scab and the rate of re-epithelialisation in the skin of the young domestic pig. Nature 1962;193:293–4.
[4] A prescriber's guide to dressings & wound management materials. VFM Unit, Welsh Office Health Department; 1997.
[5] Bennett G, Moody M. Wound care for health professionals. London: Chapman and Hall; 1995.
[6] Dealey C. The care of wounds. Oxford: Blackwell Science Ltd; 1994.
[7] Leaper DJ, Harding KG, editors. Wounds: biology and management. Oxford: Oxford University Press; 1998.
[8] Thomas S. Wound management and dressings. London: The Pharmaceutical Press; 1990.
[9] Wardrope J, Smith JAR. The management of wounds and burns. Oxford: Oxford University Press; 1992.

[10] Beldon P. Recognising wound infection. Nurs Times 2001;97:3–4.
[11] Hunter JE, Teot L, Horch R, et al. Evidence-based medicine: vacuum-assisted closure in wound care management. Int Wound J 2007;4:256–69.
[12] Berry DP, Harding KG, Stanton MR, et al. Human wound contraction: collagen organization, fibroblasts, and myofibroblasts. Plast Reconstr Surg 1998;102:124–34.
[13] Morgan DA. Wounds: what should a dressing formulary include? Hosp Pharm 2002;9:261–6.
[14] Lazarus GS, Cooper DM, Knighton DR, et al. Definitions and guidelines for assessment of wounds and evaluation of healing. Arch Dermatol 1994;130:489–93.
[15] Percival JN. Classification of wounds and their management. Surgery 2002;20:114–7.
[16] Harding KG, Morris HL, Patel GK. Science, medicine and the future: healing chronic wounds. Br Med J 2002;324:160–3.
[17] Moore K, McCallion R, Searle RJ, et al. Prediction and monitoring the therapeutic response of chronic dermal wounds. Int Wound J 2006;3:89–96.
[18] Krasner D, Kennedy KL, Rolstad BS, et al. The ABCs of wound care dressings. Wound Manage 1993;66:68–9.
[19] Russell L. Understanding physiology of wound healing and how dressings help. Br J Nurs 1999;9:10–21.
[20] Rothe M, Falanga V. Growth factors, their biology and promise in dermatologic disease and tissue repair. Arch Dermatol 1989;125:1390–8.
[21] Shakespeare P. Burn wound healing and skin substitutes. Burns 2001;27:517–22.
[22] Cooper DM. Wound healing: new understandings. Nurs Pract Forum 1999;10:74–86.
[23] Queen D, Orsted H, Sanada H, et al. A dressing history. Int Wound J 2004;1:59–77.
[24] Boateng JS, Matthews KH, Stevens HNE, et al. Wound healing dressings and drug delivery systems: a review. J Pharm Sci 2008;97:2892–923.
[25] Agu O, Baker D, Seifalian AM. Effect of graduated compression stockings on limb oxygenation and venous function during exercise in patients with venous insufficiency. J Vasc Surg 2004;12:69–76.
[26] Berliner E, Ozbilgin B, Zarin D. A systematic review of pneumatic compression for treatment of chronic venous insufficiency and venous ulcers. J Vasc Surg 2003;37:539–44.
[27] Mariani F. Compression. Torino: Edizioni Minerva Medica; 2009.

Medical bandages and stockings

8

8.1 Introduction

Edema is a condition of abnormally large fluid volume in the circulatory system or in tissues between the body's cells. For a healthy human body, the venous and lymphatic systems collaborate to maintain the micro- and macro-circulatory homeostasis of the tissues. Ulceration of the skin develops when the normal functions of these two systems are disturbed either as a result of diabetic diseases or by the application of excessive pressure on the tissue. This leads to large amounts of edema and wound exudate, with peri-wound skin having dermatitis, hyperemia or maceration, dark pigmentation, atrophy blanche, and fibrosis. Figure 8.1 shows an example of a patient with venous leg ulcers.

Clinically, a limb edema is the result of swelling from excessive accumulation of serous fluid in tissue. It depends on two factors. First, excessive liquid is generated in the interstitial space due to obstructive venous diseases presenting as varicose veins or postthrombotic syndrome, heart or kidney failure, metabolic deficits with altered and low protein concentration in the blood, or many other conditions. Second, a lymphatic drainage deficit due to alteration of the lymphatic system can result in water and protein stagnation. Lymphedema frequently occurs in many upper- and lower-limb pathologies.

From a chemical-physical point of view, the high concentration of the stagnating fluid's protein content, sometimes in excess of 1 g/dL, is the main cause of lymphedema. These above-normal proteins draw water into the venous system, thus initiating and prolonging edema. While the water fraction can be removed by diuretics or pressotherapy, the result is only temporary due to the large amount of proteins still in the interstitial space. Over the years it has been found that the two most effective therapeutic methods for the recovery of lymphostasis protein components are elastic-compressive bandage combined with manual lymphatic drainage. This basic concept highlights the usefulness of elastic compression in lymphatic stasis.

Textile-based materials such as elastic bandages and stockings can be fitted or enclosed on body parts. Bandages are used for directed compression and/or safeguarding of function, such as the stabilization and control of movement of body parts or single joints. In addition, elastic bandages can help fix wound dressings (fixation bandages), compress body parts (compression bandages), and immobilize body parts (suspension/support bandages). Similarly, compression stockings can be used as an appliance to give compression impact on the venous or lymphatic vascular system during complex physical antiedema therapy. Other forms of textiles can also be used, such as compression tubes for arms and legs, and a compression bra.

Medical Textile Materials. http://dx.doi.org/10.1016/B978-0-08-100618-4.00008-X

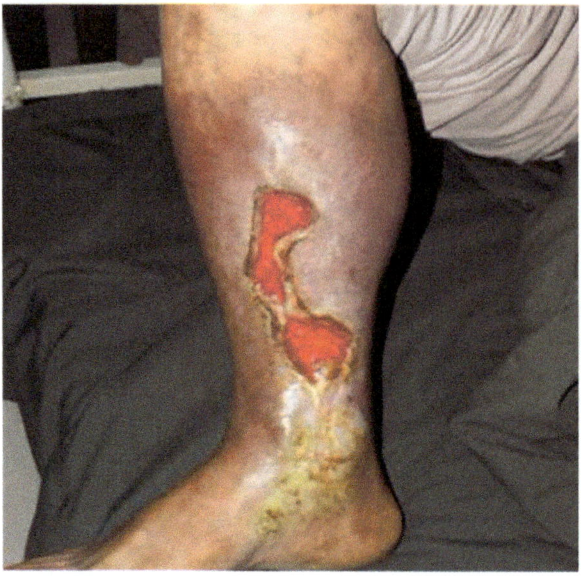

Figure 8.1 An example of venous leg ulcers.

8.2 Venous and lymphatic diseases

Common venous and lymphatic diseases include chronic venous insufficiency (CVI), varicose veins, and lymphedema, and compression forms a part of the complex treatment program.

8.2.1 Varicose veins

The venous system of the lower limbs consists of an interconnected network of superficial veins, perforator veins, and deep veins, which serve to return blood from body tissues to the heart. Veins have one-way valves that help keep blood flowing toward the heart. If these valves are weak or damaged, blood can back up and pool in the veins. This causes the veins to swell and can lead to varicose veins, which are extended and tortuous superficial veins usually of the lower extremity.

8.2.2 Chronic venous insufficiency

Chronic venous insufficiency is characterized by chronic reflux disorder of blood from the peripheral to the central vein, with subsequent venous hypertension and resulting changes in the skin. Three stages can be used in classification of CVI, characterized as follows.

Stage 1: Edema and pigmentation.
Stage 2: Edema, skin pigmentation, and dermatitis.
Stage 3: Varicosities, edema, skin pigmentation, dermatitis, and ulceration.

A CEAP classification system is used in German-speaking countries, where C, E, A, and P represent clinical symptoms, etiology, and anatomical and pathological aspects of chronic venous insufficiency.

8.2.3 Lymphedema

A lymphedema occurs as a swelling and increasing hardening of connective tissue by an interruption of the lymph drainage, which leads to the accumulation of proteins, water, fats, and cells or cell components in the lymphatic system. There are many causes of lymphedema, such as a congenital interruption of the lymph drainage, anatomically or physiologically incorrectly developed lymph vessels, benign or malignant tumors, and cases where the lymph vessels are destroyed by injury, infection, tumor surgery, or radiation treatments. Sometimes a lymphedema located on the shoulder or upper arm is caused by a breast cancer surgery, in which the lymph nodes had to be removed from the armpits. Massive interruption of lymph drainage can cause body parts to swell dramatically, leading to a disease pattern known as elephantiasis. In practice, the development of lymphedema can be divided into four stages.

Stage 0: Latent edema. The lymphatic vessel system has already sustained damage and functions at a reduced level. It can still, however, create lymph drainage conditions. Edema is not yet present.

Stage I: Reversible edema. The lymphatic vessel system cannot cope. Soft-tissue swelling occurs, mainly in the evening. The swelling can be fully reversed overnight or after a few days with limb elevation.

Stage II: Chronic irreversible edema. The swelling has clearly hardened. The skin no longer indents when pressed and the edema cannot be reversed after longer periods of rest.

Stage III: Elephantiasis. The swelling takes on extreme dimensions. The leg is badly deformed. The skin is hard and characterized by wart-like outgrowths. Due to the development of lymph sacks and excessive bulging, there is increased danger of erysipelas.

8.3 Compression therapy

Compression therapy involves the use of external pressure on the human body to treat venous and lymphatic diseases. Because liquids are able to swerve in the vascular system, the external pressure exerted on the body can have an obvious effect upon the tissues which are high in liquid content. In a medical sense, compression pressure is the resulting pressure on the enclosed tissue and embedded blood and lymphatic vessels. It can be generated by means of compression from outside (compression stocking, bandage, etc.), or from inside by the muscle which expands against the resistance of the compression.

Compression pressure is normally stated as millimeters of a mercury column (mmHg). The value of the compression pressure for elastic textile materials can be determined according to the Laplace law in the equation $P = S/R$, where P is the compression pressure, S is the tension of the textile (in Newton per centimeter of bandage width), and R is the radius of the enclosed extremity. This means that the smaller the radius (with constant tension), the higher the pressure values become. It should be noted here that the descriptions of the physical parameters in medicine (such as pressure and tension) differ from those used in engineering sciences. There is also no standard nomenclature known in literature.

In principle there are two kinds of pressure with compression bandages: static and operating. Static pressure is the permanent pressure that a compression material exerts to tissue and vessels while the musculature is relaxed. It is often described as the applied pressure, and is related to the power to apply or stretch a compression bandage during its application. The particular value depends on the type of bandages, the pre-stretching during application, the strength of the bandage or stocking, and the state of the bandage, such as its age and number of wearing and washing cycles.

When applied on patients, the compression should release continuously in the direction of the intended liquid transport, such as from distal to proximal for the leg. This pressure gradient is achieved by applying a compression bandages with constant elongation running from the ankle to the thigh. To achieve a constant pressure distribution over the whole leg, it is essential to pad the margins and fill other parts to achieve an almost cylindrical shape. Even so, because of the anatomy, good pressure distribution is very difficult to achieve, especially on the feet. Pressure values can be comparatively high at bone margins because the bandage is more stretched at these points; meanwhile, low or even no measurable pressure values could be registered at points like the perimalleolar areas.

8.4 Effects of compression therapy for the human organism

Graduated compression bandages and stockings are a recognized method in the management of venous hypertension, prevention of leg edema, and reduction in the recurrence of venous ulcers. Compression therapy is the treatment of venous insufficiency, varicose veins, or venous ulceration of the lower limbs by the patient wearing compressing garments such as bandages and stockings. It should be noted that venous leg ulcers account for an estimated 70–90 per cent of all lower-extremity ulcers, and these diseases are typically cyclical and chronic, with periods of healing followed by recurrence. Research has shown that the recurrence rates of venous ulcers are higher in patient populations who do not wear compression stockings, while wearing compression stockings has been shown to decrease the frequency of recurrent ulcers. Overall, it has been found that compression therapy has the following clinical benefits.

8.4.1 Acceleration of venous flow

The compression of extremities such as the leg with bandages, stockings, or other means leads to the formation of a pressure gradient from superficial tissue areas to the underlying areas, leading to a 1.5-fold increase of venous flow rate. It has been found that the application of compression therapy increases the healing rate of venous leg ulcers and improves deep-vein hemodynamics.

8.4.2 Influence on arterial perfusion

The increase of external pressure on the body parts from compression increases the rate of blood flow. In addition to increasing venous return, it also affects arterial flow regulation by increasing hyperemic peak flow.

8.4.3 Central hemodynamic effects

The compression of extremities, especially the legs, leads to a significant relocation of liquids from the periphery to central body parts. This increases the preload of the heart and, in the worst cases, excess relocation of liquids can cause central hemodynamic effects and lead to cardiac problems.

8.4.4 Functional improvement of the muscle pump

In the human body, the soleus is a powerful muscle in the back part of the lower leg, i.e., the calf. It runs from just below the knee to the heel, and is involved in standing and walking. In the upright posture it is responsible for pumping venous blood back into the heart from the periphery, and is often called the skeletal-muscle pump or peripheral heart. The reduction of vein diameter by means of compression bandages helps to improve the function of the muscle pump by enhancing the efficiency of the reverse flow.

8.5 Development of medical bandages and stockings

There has been a long history of the use of bandages and stockings in the treatment of venous or lymphatic system diseases, based on the understanding that the application of textile materials as adjuvants during therapy for these diseases can lead to directed removal of liquids. In this respect, when applied on the affected body parts with preliminary tension, flexible textile structures such as bandages and stockings can maintain a defined pressure pattern which promotes liquid flow in the tissue and reduces the effect of edema. It was noted by Richard Wiseman (1622–1676), Charles II of England's sergeant surgeon, that the use of a laced stocking made of dog skin had an effect on "varicose ulcer." In the same era, the Parisian surgeon Pierre Dionis (1643–1718) described the use of elastic laces and flannel bandages, while Johann Christian Anton Theden (1714–1797), surgeon general to the court of Frederick II of Prussia, illustrated bandaging techniques with the detail of including the fingers or toes, stitching the turns, or wearing a sock to avoid displacement of the bandage in the first part of his work "The great benefit of wrapping the extremities in bandages," which appeared in 1771. Compression with India rubber was described by Wye in 1781, and soon after there were reports on the use of inflatable India-rubber leggings for ulcers or even compression packs impregnated with turpentine wax and covered with bandages of different types, such as leather, cardboard, cork, and celluloid. In 1797 Thomas Baynton published his work "New method of treating old ulcers," which described a tight bandage with plasters of pitch resin cut in strips and positioned on the limb-like tiles. This product was able to exert strong pressure to approximate the margins of the ulcer.

In 1870 Martin of Boston introduced the use of a highly extensile rubber bandage in direct contact with the skin and covered by a second bandage. Shortly afterward, in 1885, a dermatologist in Hamburg named Paul Unna marked another historic point with his rigid zinc oxide bandage (Unna's paste boot), which was the first industrial example of adhesive bandages. In 1929, in association with Beiersdorf, N. Brann produced "Novoplast" bandages which had adhesive on both sides. Karl Sigg (1912–1986), a supporter of compression therapy in sclerotherapy with his famous method, established the scientific basis of compression therapy together with the Dutch Van der Molen and the French Robert Stemmer. Karl Sigg was also the inventor of the graduated compression elastic stocking produced by the Ganzoni Sigvaris® company from 1958 to 1960.

Figure 8.2 illustrates the application of medical stockings, which were developed in the nineteenth century after the discovery of rubber vulcanization by Charles Goodyear in 1839. In 1846 the first patent for fine and quadrangular rubber thread was registered in England, and this rapidly conquered the textile industry. The elastic stocking was born on 26 October 1848 when William Brown of Middlesex registered Patent No. 12294 concerning an elastic stocking in India rubber manufactured on hand

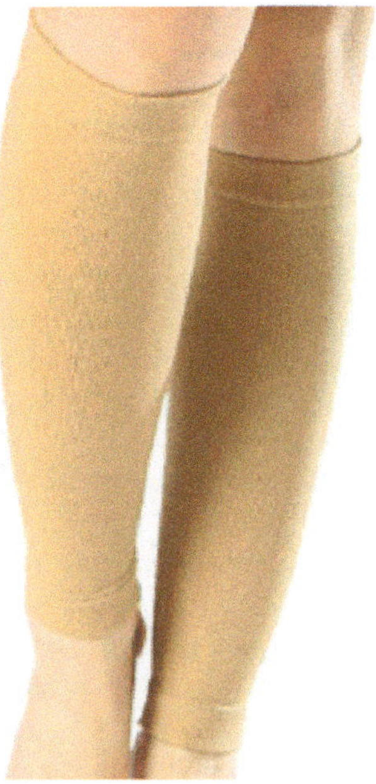

Figure 8.2 Application of medical stockings.

looms. Shortly afterward, in 1851, Jonathan Sparks introduced threads made of silk and cotton. In 1861 William Saville made the first surgical elastic stockings on hand looms. Almost concurrently the Ganzoni elastic materials factory (today Ganzoni Sigvaris®) was founded in Winterthur in Switzerland in 1864, and then became the world leader in the production of medical elastic stockings.

8.6 Classification of elastic compression bandages

Elastic bandages can be classified based on their applications and product performance. Rigid bandages are mainly used for fixation, where rigidity is defined as the ability of the bandage to oppose muscle expansion when it contracts. It depends on the elastic recoil of the material used in its construction. The more compact the weft (and with a lower presence of elastic fibers), the shorter the stretch of the bandage and the more its rigidity. The rigidity of a bandage can be measured with simple, economical, and reproducible methods and expressed with various indices, the most studied of which is the "static stiffness index," which is the difference between the pressure exerted in vivo by the bandage on the surface of the leg standing and lying. If this pressure difference is greater than 10, the bandaging is of the rigid type. Stiffness signifies the increase in the pressure of the bandage (given by the resistance offered by the bandage) with a 1 cm increase in the circumference of the limb.

Elasticity and extensibility are the two most important characteristics of compression bandages. Elasticity is defined as the ability of the bandage to regain its original form after extension, due to the addition of elastic threads in the longitudinal direction of the bandage. The force employed to stretch the bandage (deformation energy) indicates the elastic power or tension. The ability of the bandage to maintain tension and therefore the pressure it exerts depends on its elastomer properties, which in turn depend on the thread and the method used to construct the fabric. Extensibility is defined as the ability of the bandage to elongate when subjected to stretching; it is measured by the lengthening of the bandage when a force of 10 N is applied per cm of height and is expressed as a percentage of the length at rest. Once a given extension has been reached, the physical structure of the bandage prevents further stretching. This condition is called "locking." According to Stemmer's classification, short-stretch bandages should lock once a maximum of 70 per cent of extension has been reached (preferably 30–40 per cent), while long-stretch bandages should lock beyond 140 per cent.

Because of their respective long histories and tradition, the German-speaking countries adopted different standards to those used in England. The German standard (RAL-GZ 387) classifies bandaging materials into the following types.

1. Material for protection, absorption, and fixation (orthopedic wool, light cohesive bandages, foam) and material for eccentric compression.
2. Nonelastic bandages (rigid zinc oxide bandages).
3. Short-stretch bandages (40–70 per cent).
4. Medium-stretch bandages (70–140 per cent).
5. Long-stretch bandages (>140 per cent).

The English standard (BS 7505:1995) divides bandages into three types.

1. Fixation bandages (orthopedic wool, foam, light cohesive bandages).
2. Short-stretch bandages (zinc oxide bandages, short-stretch elastic bandages).
3. Long-stretch bandages.

In both classification methods there are four classes in the long-stretch elastic bandages, with different elastic strength and different resting pressure, which is calculated at a known ankle circumference (23 cm) with bandaging overlapping of 50 per cent.

Based on their adhesive properties, compression bandages can be further divided into:

- nonadhesive bandages;
- cohesive bandages;
- adhesive bandages.

Cohesive bandages are manufactured by atomization of the surface of both sides of the bandage with latex or acrylic microspheres diluted in water, which evaporates on heating to 50 °C in suitable furnaces. The latex suspension is fixed and covers the entire surface of the bandage, and is able to adhere only to itself. Acrylic and zinc oxide glues are highly hypoallergenic, and these bandages can be used relatively readily in direct contact with the skin.

Adhesive and cohesive bandages are usually short- to medium-stretch bandages which are able to stick to the skin and to themselves (adhesive bandages) or only to themselves (cohesive bandages) by means of acrylic adhesive or zinc oxide glue. Technically, adhesive bandages can be manufactured using elastomers (natural rubber, etc.) or be made completely of cotton with a defined elastic weave. Some adhesive elastic bandages are made of pure cotton without elastomers and with a particular weft, producing elasticity in the bandage which is defined as "variable memory." These bandages, which often require shaving when they are placed in contact with the skin, have the adhesive arranged in strips with a protective film which is removed when they are used. If they are stretched to up to 50 per cent of their length (submaximal tension), they behave like normal elastic bandages and can regain their original length, but if the stretch exceeds 50 per cent they lose their elastic capacity (maximum tension) and are transformed into rigid bandages. These bandages are used mainly in orthopedics, sports medicine, and physiotherapy.

During application it is preferable to use a skin protector (polyurethane foam film with high porosity) beneath the adhesive bandage even if this technique may reduce the stability of the bandage (slipping). The use of elastic tubing beneath the bandage produces greater stability of the bandage itself with the addition of the pressure exerted by the tubing, which is about 10 mmHg at the ankle. Adhesive and cohesive bandages are permanent (fixed bandages that the patient must not remove), which are more stable over time and provide medium-to-high pressures on the tissues. This type of bandage is indicated when the patient must leave the bandage on both day and night for a long period, such as in the treatment of venous ulcers, edema, postoperative, or postsclerotherapy.

8.7 Elastic stockings

Elastic stockings are elasticated hose which can apply graded compression to lower-leg soft tissues in order to reduce edema and prevent venous clot formation. In clinical trials and world scientific literature, the therapeutic elastic stocking has demonstrated a certain efficacy in the prevention and treatment of deep-vein thrombosis. To ensure its performance as a medical device, several national standards, such as the German RAL-GZ 387, French NFG 30-102B IFTH, and English BS-7505, specify manufacturing methods which guarantee a defined and graduated pressure along the limb between certain parameters. These standards differ only in the compression range of various therapeutic classes, for which an attempt is being made to achieve uniformity in Europe. In general, the standards specify the following requirements.

1. Manufacture with circular looms (or linear for some types of made-to-measure stockings).
2. Production with materials of a quality and using methods defined in the standard.
3. Graduated and uniform compression from below to above.
4. Declaration of the compression class in mmHg at the ankle and along all the lower limb.
5. Extensibility of the knit in both directions to facilitate joint movement.
6. Aerated fabric to allow evaporation and cutaneous transpiration.
7. Knitted heel to allow anatomical position of the ankle and therefore the required compression in the specific region of the leg.
8. Perfect adaptability to the shape of the lower limb.
9. Guarantee of the duration of compression for a period of at least 4–6 months, depending on the type of threads used in manufacturing the stocking.
10. Production and product expiry dates on the packaging together with all the details needed for certifying traceability and the quality of the medical device.
11. Inspection of the products with tests conducted constantly by independent authorized institutes.

Technically, the elastic stocking consists essentially of a transverse weft thread which produces the compression. The weft thread is made of natural rubber or elastane (Lycra fibers), which is usually double wrapped with threads such as cotton, polyamide, or microfiber polyamide so as to fix the weft thread itself at a predetermined tension and produce the physical and esthetic characteristics of the stocking. The knit through which the weft thread passes consists of elastic yarns, of lower strength compared with the weft, usually woven with the method known as crossed and locked knit.

8.8 Characteristics of textile compression materials

Textile-based compression materials include elastic bandages, stockings, and garments that are designed to exert external pressure to the human body. The main product performance is the force-elongation behavior, which is related to the fabric structure, yarn composition, and fiber type. For elastic compression materials, the textile structures are stretched during application, and the resulting tension creates a compression pressure on the underlying body. During wear, the compression materials alternate between static and operating pressure according to the patient movements.

In addition to the force-elongation behavior, elasticity is an important property for compression materials, and is closely related to fatigue of the compression materials during use. Elasticity characterizes a material's ability to elongate under the impact of a tension force and return all or part to its original dimension when the tension is released. Based on their elastic properties, bandages are broadly divided into two groups, rigid or nonelastic and elastic types, with the latter further divided into short stretch (<70 per cent of the original length), medium stretch (70–140 per cent), and long stretch (>140 per cent) according to the type of elastomers used for making the product. An example of a rigid or inextensible bandage is the zinc oxide bandage conceived by the dermatologist Paul Unna in 1885. Other types of inextensible bandages are usually made of cotton blended with 20–40 per cent polyamide fibers. For short-stretch bandages, the optimal elasticity is about 30–40 per cent, which is required to obtain appropriate working pressure on the body while at the same time avoiding the disadvantages of a completely rigid bandage, such as the difficulty of applying the bandage and keeping it in place over time. For medium- and long-stretch bandages, a distinction should be made between one-way or two-way stretch bandages according to the direction of stretch. A high level of elasticity is achieved by using natural or synthetic elastic threads of various types, such as natural rubber, polyamide (nylon), and elastane (Lycra), in combination with traditional fibers such as cotton. Since the middle of the twentieth century synthetic rubbers have replaced natural rubber in the manufacture of elastic threads, while polyamide fibers such as nylon 6, nylon 66, and nylon 12 are also widely used, thanks to their ability to offer a high resistance to traction, solvents, and other ingredients associated with elastic bandages.

During actual use, bandages lose about 40 per cent of their original pressure after 120 min, due to the volume reduction of the bandaged limb and the relaxation of the bandage structure. In this respect, elasticity of the bandage is important in order to maintain a sustained pressure after the removal of a deforming force. Because of the movement of body parts, the working pressures are intermittently elevated in accordance with muscle contraction, while the resting pressures are usually lower than working pressures. Highly elastic fibers such as polyurethane-based elastic materials offer excellent performance characteristics in the development of elastic bandages and stockings. For example, Lycra fibers, which contain at least 85 per cent polyurethane by mass, are on the market in both monofilament and multifilament form, with the former used in bandages while the latter are employed in the stocking sector. These elastic threads can be stretched sevenfold and still be able to return to their original sizes.

In addition to the physical and mechanical properties, ecological safety and good wear comfort are important for patient acceptance and the desired success of therapy. In practice, compression bandages made of a large variety of materials are applied on patients in many different clinical circumstances, and a careful selection of appropriate products is needed before the application of bandages or stockings on patients.

8.9 Properties of textile compression materials

8.9.1 Compression force

During their application, textile compression materials are stretched and wrapped on to patients under tension, which generates a compression force during wearing. This force is dependent on a number of factors, such as the chemical structure of the elastomer, the texture of the textile fabric, the characteristics of the yarn and thread, the crossing of threads during weaving, knitting, or mesh formation. Since each patient has his or her individual body geometry, in order to accomplish a desired therapeutic success it is necessary to adapt the compression bandage or stocking to the individual's body geometry, so that the required pressure behavior is realized. At the beginning of therapy the size and length dimensions are first gathered with a measuring tape. A fitting compression stocking can then be selected using the collected data according to the size measurement table of a valid standard.

8.9.2 Elongation

The elongation of compression bandages is determined in a tension test, with a tensile load of 10 N/cm of the bandage's width and as used in praxis with the normal load of 3 N/cm of the bandage's width. Figure 8.3 shows testing for the elongational properties of a compression bandage.

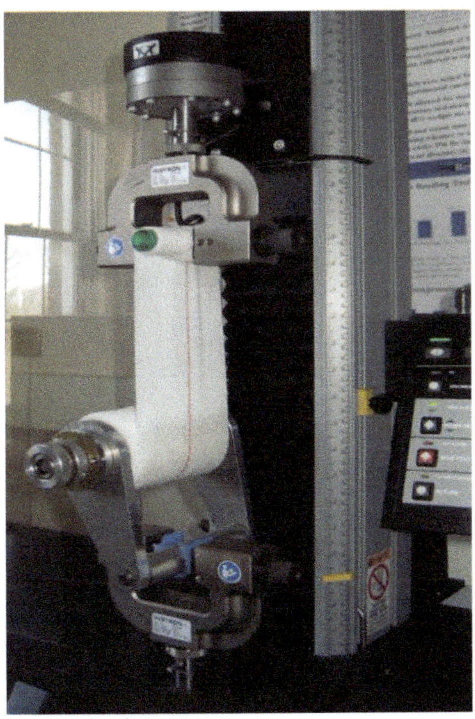

Figure 8.3 Testing for elongational properties.

8.9.3 Working and resting pressure of bandages

Concerning the pressures generated by textile compressions bandages, working pressures refer to those under muscle contraction while resting pressures are obtained during muscle relaxation or rest. Different types of bandages have considerably different pressure profiles during application. Nonelastic (rigid) bandages create peak pressures, and conversely elastic bandages, especially long-stretch ones, absorb the volumetric changes of muscle contraction in the limb and do not demonstrate peak pressure variations but more constant pressure, with minimal pressure changes during walking. When applied, the more a bandage is able to give peak pressures (working pressures), the more the action is transmitted deeply (into the deep-venous system). Within certain limits, a long-stretch bandage can be transformed into a medium-stretch bandage and a medium- to a short-stretch bandage by doubling the number of turns of the bandage, where the overlapping of the turns causes not only an increase in the pressure exerted but also a shift of the system toward rigidity.

8.9.4 Elasticity

Elastic compression stockings feature elasticity in both longitudinal and crossways directions because of knitwear's structure. Elasticity in cross-direction plays an important role in the application of the stockings when pulling them with the lowest diameter over a significantly bigger size of heel and instep.

8.9.5 Wearing comfort

Wearing comfort is a combination of thermo-physiological wearing comfort, skin-sensory properties, and ergonomic wearing comfort, which make the usage as comfortable and adequate as possible. It is achieved through proper heat and moisture exchange in combination with skin-sensory properties of the product, such as haptics and softness.

8.9.6 Pressure measuring systems

Because of the variations in the sizes of patients' legs and the different techniques adopted by different therapists, the pressure generated from compression differs even when the same type of bandage is applied on the same patient. In this respect, direct recording of compression values between the bandage and the skin is the best method for assessing the technical performance of compression materials. Thus although compression pressure can be tested in an indirect in vitro method, it is more appropriate to carry out direct in vivo tests. The goal of the direct in vivo method is the measurement of compression pressure directly on the affected extremity. The problem here is that human extremities have no ideal cylindrical shape and the compression pressure is inversely proportional to the radius of the extremity. Thus the compression pressure can only be determined on a limited and small area with a constant size. This means that the sensor itself influences the measuring values through its construction height.

Therefore the results of these methods are only approximated values for defined areas of the leg.

Over the years many direct in vivo methods have been developed, including the following.

- Tonometer by Van der Molen, introduced in 1955.
- SIGG-Tester, introduced in 1964.
- Measuring tool by Assum, introduced in 1965.
- Komprimeter by Haid, introduced in 1965.
- Medical stocking tester, described in 1978 for the first time by Borgnis de Bruyne and Dvorak.
- Talley system, introduced in 1993.
- MCDM contact pressure system by Klyscz, introduced in 1996.
- Contact pressure measuring system by Blazek and Schultz-Ehrenburg, introduced in 1997.
- Novel-pliance pressure measuring system, available since the 1990s in different layouts.
- Kikuhime pressure measuring system, developed at the end of the 1990s at the Technical College Hokkaido, Japan.

In recent years so-called "evidence-based medicine" has been used as a practical tool to assess the clinical efficacy of medicines and medical devices. With compression products, evidence-based compression therapy serves as a quality control procedure and helps to deliver the efficiency statements which service providers require for payment for the accordant therapy services. In practice, compression products should be subjected to appraisal by randomized and monitored studies which investigate the compression therapy process and produce proof of effectiveness. An important objective of this investigation is the resulting contact pressure following the application of the compression product. Attempts should be made to carry out direct in vivo determination of the compression pressure or the chronological pressure profile on patients.

8.10 Summary

The development, production, and application of medical compression textiles are a highly complicated process. On the one hand, medical and therapeutic requirements such as compression pressure and pressure distribution need to be fulfilled by the use of appropriate textile technologies. On the other hand, the individual adaptation of compression textiles concerns modular bandaging systems, bandaging materials, customized design and manufacturing, and the application of pressure-measuring technologies. The common goal of all these different aspects of medical compression materials is an optimal adaptation of the compression textiles to the individual therapeutic and anatomical situations. This should be made easier with advances in textile technologies and computer sciences, which will enable the customization of compression stockings and bandages that are manufactured individually for different body geometries with programmable knitting machines.

Sources for further information

[1] Mariani F. Compression. Torino: Edizioni Minerva Medica; 2009.
[2] Bartels VT, editor. Handbook of medical textiles. Cambridge: Woodhead Publishing; 2011.
[3] Lai CH, Li-Tsang CW. Validation of the Pliance X System in measuring interface pressure generated by pressure garment. Burns 2009;35:845–51.
[4] Partsch H. Die Phasengerechte Kompressionsbehandlung beim Ulcus cruris venosum. Orthopadie-Technik 4:238–245.
[5] Partsch H. The static stiffness index: a simple method to assess the elastic property of compression material in vivo. Dermatol Surg 2005;31(6):625–30.
[6] Agu O, Baker D, Seifalian AM. Effect of graduated compression stockings on limb oxygenation and venous function during exercise in patients with venous insufficiency. J Vasc Surg 2004;12:69–76.
[7] Benigni J, Sadoun S, Allaert F. Efficacy of Class 1 elastic compression stockings in the early stages of chronic venous disease. Int Angiolgy 2003;24(3):224–38.
[8] Berliner E, Ozbilgin B, Zarin D. A systematic review of pneumatic compression for treatment of chronic venous insufficiency and venous ulcers. J Vasc Surg 2003;37:539–44.
[9] Buchtemann AS, Stein A, Volkert B. The effects of compression therapy on venous hemodynamics in pregnant women. Br J Obstet Gynaecol 106:563–69.
[10] Choucair M, Phillips TJ. Compression therapy. Dermatol Surg 1998;24(1):141–8.
[11] Ibeguna V, Delis KT, Nicolaides AN, et al. Effect of elastic compression stockings on venous hemodynamics during walking. J Vasc Surg 2003;37:420–5.
[12] O'Meara S, Cullum N, Nelson E. Compression for venous leg ulcers. Cochrane Collab 2008;2:1–124.
[13] Prandoni P, Lensing A, Prins M. Below knee elastic compression stockings reduced development of post-thrombotic syndrome in proximal deep venous thrombosis. Evidence Based Med 2005;10:22–8.
[14] Ruckley C. Socioeconomic impact of chronic venous insufficiency and leg ulcers. Angiology 1997;1:67–9.
[15] Spence R, Cahall E. Inelastic versus elastic leg compression in chronic venous insufficiency: a comparison of limb size and venous hemodynamics. J Vasc Surg 1996;44:783–7.
[16] White JV, Katz ML, Cisek P, et al. Venous outflow of the leg: anatomy and physiology of the plantar venous plexus. J Vasc Surg 1996;24:819–24.

Surgical sutures

9

9.1 Introduction

Surgery is as old as humankind, and the use of sutures dates back to the times of the earliest human knowledge in the field of surgery. Historically, many diverse types of materials have been used for closing and suturing surgical wounds. The most ancient suture materials include animal tendons and hairs, as well as herbal fibers such as linen, hemp, and different grass species. Egyptian papers dating back 5000 years show scenes of suturing wounds with linen fibers. Nowadays, sutures are widely used in surgery and for closing minor or major lacerations following trauma. They are among one of the largest groups of biomaterials used as implants in the human body.

The primary function of sutures is to hold a wound in a good position for the promotion of wound healing. Incised surgical wounds and clean, fresh incised traumatic wounds which are not associated with contamination or devitalized tissue can be closed with sutures, staples, or tissue adhesives. Careful apposition of tissues allows healing by primary intention, resulting in cosmetically and functionally acceptable scars in skin, nerve, or tendon. Sutures are also used in bowel anastomosis and vascular arterial anastomosis, which demand a leak-proof technique using a permanent suture.

In the first few days of healing, wounds have little intrinsic strength—indeed, tissue strength is lost within 0.5 cm of the wound edges due to the action of collagenases. Adequate support of the wound with sutures placed beyond this area of tissue reaction is essential during this period. Gradually, the tensile strength of the healing tissue increases and suture support becomes less important. Different tissues heal at different rates, but even a year or more after wounding, most wounds will only have about 70% of the tensile strength of unwounded tissue. In addition, several factors can delay healing, such as infection, hematoma, steroids, and coexisting disease causing immune incompetence, malnourishment, and cachexia. Tissues that heal rapidly and are not under mechanical stress can be sutured with absorbable sutures, such as in gastrointestinal surgery. Tissues exposed to disrupting forces during healing or which heal slowly, such as in cardiac and vascular surgery and abdominal wall closure, need prolonged support and generally require suturing with nonabsorbable suture material. Nonabsorbable sutures should be avoided in bile duct and ureteric surgery because they can act as a nidus for stone formation. Skin sutures need to be left in situ for different periods of time, depending on the area of the body that has been sutured. The skin of the face and neck heals rapidly and sutures can be removed after 4–6 days. In this short time "tram-line" scars will not develop and therefore sutures may be placed in an interrupted fashion and still achieve excellent cosmetic results. Most

Medical Textile Materials. http://dx.doi.org/10.1016/B978-0-08-100618-4.00009-1

of the skin elsewhere on the body requires suture support for 8–14 days to achieve adequate wound healing. Sutures that are removed after this length of time may leave "tram-line" scars unless a subcuticular technique is used. Skin may be sutured with absorbable materials, thereby avoiding the need for suture removal.

Sutures are foreign bodies, and although modern synthetic monofilament sutures are almost completely inert, braided and "natural" sutures can excite a significant inflammatory response. Bacteria can be difficult or impossible to eradicate once colonization of the suture or knot interstices has occurred. This can predispose the wound site to infection, with abscess and sinus formation, wound dehiscence, and incisional hernias. Sutures should be avoided or used with caution in the presence of established infection or severe contamination.

It should be pointed out that in addition to sutures, mechanical devices such as clips and staples are also used to close open wounds. Skin wounds may be closed with clips or staples, typically after thyroid surgery or following vascular or orthopedic surgery, which tends to involve long wounds. These devices give a good cosmetic result and are quick and easy to use, but are considerably more expensive than the equivalent length of suture. However, the savings in operating theater time more than make up for the extra expense. In addition to clips and staples, alternative wound closure devices on the market include closure strips, tissue adhesives, and sealants, although these are of insufficient strength for most surgical needs.

9.2 Historical development of surgical sutures

Sutures are used for closing wounds in all body tissues and are the most frequently implanted foreign materials. Of the estimated 50–60 million open surgical procedures performed each year, one of the common challenges is the need to connect tissues. This typically involves closing a wound or incision across various levels of muscle, fascia, fat, and skin.

The history of suture development details the use of a variety of naturally occurring materials. Evidence of suturing wounds dates back to between 5000 and 3000 BCE with the invention of eyed needles made of bone. Early references to suture materials were made in the Caraka Samhita, written 3000 years ago, and the Sushruta Samhita, written 2500 years ago, in which the use of plaited horsehair, cotton, leather strips, and tree-bark fibers for closing wounds was described. The Edwin Smith papyrus details Egyptian wound-closure techniques dating back to 1600 BCE. The treatment of a gash was described as "you should draw together for him his gash with stitching." Linen strips coated with resin or an adhesive mixture of honey and flour appeared in ancient Egyptian medicine, as forerunners of our modern-day skin closure tapes. In ancient Rome Cornelius Celsus described the use of small metal clips, similar to the Michel clips of today. Galen repaired the severed tendons of gladiators with sutures, giving some hope for recovery of function where previously there had been only paralysis. He used linen, silk, and catgut sutures. Avicenna, in about 920 CE, realized that in the presence of gross infection, traditional materials such as linen thread broke down rapidly. He used pigs' bristles as an alternative. Gold wire sutures were introduced by Fabricius in 1550 and silver wire by Marion Sims in 1850. By the eighteenth century silk sutures had gained popularity, along with waxed thread, although catgut was still probably the material most often used. In the nineteenth century Joseph Lister stored catgut in a solution of olive oil and carbolic acid. He chromicized catgut to improve its handling characteristics, and found the unexpected additional benefit of delayed suture absorption.

Since the mid-twentieth century technological advances have produced a range of synthetic absorbable and nonabsorbable materials for use as sutures, in braided and monofilament forms. Both synthetic and natural sutures benefited from advances made in the textile industry which allow braiding of suture material and coating of sutures with a variety of agents, such as wax, liquid paraffin, silicone, and polytetrafluoroethylene to improve handling and knotting capabilities and also reduce the infective potential through the capillarity of suture interstices. At the same time, the needles used with sutures have also improved, with the wide use of stainless steel needles. The shape of needle tips has been fashioned to suit their intended purpose, for example triangular-tipped (cutting) needles for easy penetration, and round-bodied needles for minimal trauma. A wide range of needle sizes and shapes is now available, with needles for microvascular work being barely visible to the naked eye.

In terms of the materials used for making surgical sutures, there have been substantial changes from natural materials such as catgut, cotton, and surgical silk to more advanced synthetic materials. Nylon and polyester were among the first synthetic suture materials, produced in the 1940s. Soon after, polyethylene and polypropylene were developed. In 1970 polyglycolic acid was developed under the licensed name Dexon.

9.3 Characteristics of surgical sutures

During the production process, sutures are carefully designed to give the optimum combination of strength, uniformity, and hand. The term "hand" is the most subtle of all suture quality aspects. It relates to the feel of the suture in the surgeon's hands, the smoothness with which it passes through tissue and ties down, the way in which knots can be set and snugged down, and most of all the firmness of the suture. Extensibility relates to the way in which the suture will stretch slightly during knot tying and then recover. The stretching characteristics provide the signal that alerts the surgeon to the precise moment when the suture knot is snug.

When used on wounds, suture materials are characterized by four basic features: knot safety, stretch capacity, tissue reactivity, and wound safety. Knot safety implies the capacity of suture material to retain the strength and firmness of the knot without slackening over time. It is inversely proportional to a suture thickness, and directly proportional to its quality and functional output. Stretch capacity of the suture material is defined as a stretch power capacity per unit of area. Wound safety is associated with a suture's ability to ensure safe healing of the wound, while tissue reaction is reflected in an inflammatory reaction which develops during the first 2–7 days after stitch application into the tissue.

Technically, surgical sutures should meet the following requirements.

1. Sufficient mechanical strength to hold the wound site, with the mechanical properties not significantly affected by body fluid before the healing of the wound.
2. High flexibility and good knot strength to allow easy operation.
3. Good biocompatibility, with no tissue reaction and no impact on tissue growth.
4. Complete biodegradability, with no foreign residues in the body for bioabsorbable types of suture material.
5. Good stability and easy to be sterilized without change of technical performance.
6. Nontoxic, with no allergic responses or other side effects.

Sutures are usually available in presterilized individual packets. They can be either mounted or nonmounted, that is, with or without needles. Needles can be cutting, reverse-cutting, or round-body, and the sizes vary.

Sutures can be single-stranded monofilament or multifilament with many strands. The United States Pharmacopoeia recommends that suture boxes provide complete information about the suture materials, including:

- size;
- types of sutures;
- types of needles.

Some boxes have specific indications, such as shown below.

FS	For skin
PS	For plastic skin
P	For precision point
PC	Precision cosmetic

In addition, as a general rule the following sizes are commonly used for suturing.

Face	5/0 or 6/0
Hands and limbs	3/0 or 4/0
Elsewhere	2/0 or 3/0

The surface characteristics of the suture material have a direct impact on clinical performance, since when implanted into the body sutures make direct contact with the tissue. In recent years bi-directional barbed sutures have been fabricated from monofilament fibers via a micro-machining technique, which escarps barbs into the fiber around the circumference in a variety of configurations. As shown in Figure 9.1, the barbs are positioned in a helical pattern and divided into two groups of varying lengths which face each other in opposing directions from the suture midpoint. Since barbed sutures self-anchor, the various challenges associated with surgical knots can be avoided. During application the lack of constrictive suture loops enhances healing in several ways, including increased blood flow, reduced scarring, and elimination of suture marks on the skin. Additionally, barbed sutures permit tissues to be approximated and held taut during suturing. This is especially advantageous in closing long

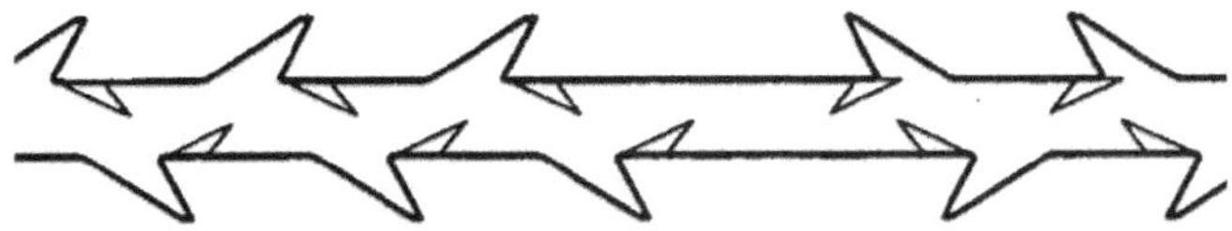

Figure 9.1 A schematic illustration of a bi-directional barbed suture.

incisions and obviates the need for a "third hand." As swelling ensues, conventional running sutures slide to equalize tension—allowing the center of the wound to widen with resultant hernias, unsightly scars, and even dehiscence. Since the use of barbed sutures can eliminate knot tying, operating time can also be reduced significantly.

9.4 Classification of surgical sutures

Depending on their chemical and physical characteristics, sutures can be divided into the following classes.

1. Absorbable or nonabsorbable.
2. Natural or synthetic.
3. Monofilament or multifilament.
4. Twisted or braided.
5. Dyed or undyed.
6. Coated or uncoated.

Ideally, sutures should disappear when their job is done. However, variable and often rapid loss of tensile strength in absorbable suture materials makes them an unreliable choice when suturing tissues which are subjected to significant mechanical stress, such as abdominal wall closure, or in situations in which premature suture failure would have immediate and catastrophic consequences, such as vascular surgery. The majority of acute wound failures are due to the "cutting out" of intact sutures from tissues, rather than suture or knot failure, and this can result following the use of any suture type. However, in general absorbable materials excite a greater tissue reaction than nonabsorbable sutures, and this augmented inflammatory response may weaken the integrity of the tissue bite. While most cases of "burst abdomen" are probably due to cutting out of intact sutures, premature suture failure can be responsible for late wound weakness and herniation.

Monofilament sutures have no interstices in which to lodge bacteria, but they are more difficult to handle and require more throws to each knot for security than braided sutures. Multifilament sutures are generally easier to handle, with greater knot security, but do allow capillarity and interstitial bacteria colonization which can increase the risk of wound infection, particularly with nonabsorbable sutures. Both monofilament and multifilament sutures can be absorbable or nonabsorbable.

Sutures are dyed in a variety of colors to enhance their visibility during surgery. Dyes do not affect the absorption characteristics of the suture. Sutures are coated or treated with a variety of agents to improve general handling and passage through the tissues, to occlude suture interstices, and to retard absorption, for example, chrome tanning of catgut. To solve the problem of nosocomial infections associated with braided sutures, antibiotic coating is applied to the sutures to release antimicrobial ingredients locally during the first few days following surgery. Two broad-spectrum antibiotics with superior gram-positive activity, moxifloxacin and clindamycin, have been incorporated into sutures. In vitro experimental results from zone of inhibition tests demonstrated that both antibiotics provided effective prophylactic bacteriostatic properties for periods of up to 2 days.

9.5 Needles

Surgical sutures are often used with needles of appropriate size and shape. Surgical needles may be curved or straight and, depending on size, may be used with needle holders or as "hand needles." Hand needles risk "needle-stick" injury and their use should be discouraged, particularly when there is a risk of viral transmission. There is a variety of different curvatures available, from a quarter of a circle through three-eighths, half, and five-eighths of a circle to "J"-needles. Needles may be "eyed," requiring the suture to be threaded, or "atraumatic," with the needle swaged on to the suture.

Needle points and cross-sections can be variable. They may be "round-bodied" in cross-section and only sharp at the tip, used mainly for suturing bowel, fat, peritoneum, and other soft tissues where the needles separate rather than cut the tissue. "Taper-cut" needles also have a round body but are sharpened on several sides toward the tip, keeping the needle tract size to a minimum. These are particularly useful in cardiac and vascular procedures. "Cutting needles" have a variety of point shapes designed for different surgical uses; they may be flattened and "sword-like," or triangular in cross-section with the apex cutting edge on the inside of the needle curvature (reverse cutting). Cutting needles are designed for ease of penetration of tough or dense tissues, including skin. They cut rather than separate tissues, and can therefore cause local bleeding by cutting blood vessels. Other modifications include blunt-point suture needles for suturing friable tissues, for example, the liver, and "heavy" needles, which are particularly strong and bear little resemblance to the size of the suture material.

9.6 Production of surgical sutures

Nowadays a large number of natural and synthetic polymers are used to manufacture suture materials, and there are many types of surgical sutures with a wide variety of performance characteristics. For absorbable sutures, the commonly used raw materials include catgut, polyglycolide, polylactide, copolymers of polyglycolide and polylactide, etc., while nonabsorbable sutures can be made from polypropylene, polyamide, polyester, and silk. Biodegradable sutures are mainly used for internal wound closures, while external wounds are generally closed with nonbiodegradable sutures which are removed when healing is started.

During the production process, either monofilament or multifilament threads can be prepared in a spinning process. Suture performance is strongly affected by material types and production methods. For example, raw material quality and production technology have an important effects on suture elasticity. If sutures are produced from the same raw material by different manufacturing methods, the mechanical properties can also change. Monofilament sutures produced from silk, polyamide 6, polyester, or polypropylene have a lower mechanical resistance than cross-braided sutures produced from the same raw materials. In addition, suture diameter also has a significant effect on its tensile strength.

Sutures are sterilized by manufacturers using ethylene oxide or gamma irradiation. Steel sutures may be sterilized by autoclave, as can those made of linen, nylon, polyester, and polypropylene.

9.7 Applications of surgical sutures

During the application of surgical sutures, the following principles of suturing should be observed.

1. Tissues should be handled gently and wound edges protected from contamination, drying, abrasion, or crushing.
2. Hemostasis should be meticulous, but beware of overuse of diathermy, particularly at the skin edges, as burning of the skin may occur, causing both a delay in healing and an increase in the size of the scar.
3. Contaminated and devitalized tissue should be excised. If severe contamination exists then the wound should not be sutured, but left open to heal by secondary intention.
4. Choose the correct size and type of suture for the task, bearing in mind the healing characteristics of the tissue in question and the tensile strengths involved. Do not use sutures that are larger than necessary, and avoid implanting nonabsorbable materials where possible.
5. Sutures should be tied tightly enough to approximate the wound edges without "strangling" the tissues. Wound edges become swollen in the early phase of healing. This will exacerbate ischemic problems caused by overtight sutures, and may result in tissue necrosis and "cutting out" of the sutures with wound breakdown.
6. Surgeons' knots should always be used, with particular regard to "laying" each throw of the knot in the correct orientation to prevent breakage and slippage of the suture. Synthetic monofilament sutures are more difficult to handle than multifilament and natural sutures, and require a greater number of throws to each knot to ensure security. Large knots, if close to the skin, should be "buried" to prevent irritation.
7. Suture materials should never be gripped with instruments, as this weakens them and can lead to suture breakage.
8. Always use appropriate lengths of suture with adequate bites to do the job. Mass closure of the abdominal wall demands suture bites of no less than 1 cm from the wound edge and about 1 cm apart. This will require a total suture length of at least four times the length of the wound.
9. Attention should be directed to appropriate inversion or eversion of the sutured tissue edges to facilitate healing and prevent complications.

10. Once the wound has been sutured, it remains at risk of becoming infected and should in general be covered with an appropriate dressing and inspected regularly until healing is sufficiently advanced. However, inspection should not disturb the "seal" of the wound, otherwise infection may be introduced.
11. Adequate instruction should be given regarding the timing of the suture removal.

In practice, much larger sutures are usually used than are strictly necessary in relation to the breaking strength of the tissues. Smaller sutures have smaller needles, which may not allow adequate tissue "bites" to be taken. They are also more difficult to handle and often more expensive than larger, more popular sutures. However, in some situations, notably ophthalmology, microvascular surgery, and plastic surgery, fine sutures are routinely employed.

During the surgical operation, many methods can be used to suture a surgical wound, including the following.

- Simple interrupted suture.
- Vertical mattress.
- Horizontal mattress.
- Subcuticular.
- Continuous.

Suturing must be carried under strict aseptic technique with the help of an assistant. First and foremost, it is important to assess and clean the wound thoroughly. The requisites of suturing include a surgical drape, local anesthesia, gloves, needle holder, dissecting forceps, scissors, cleaning lotions, and dressing materials. It is also important to observe clinical complications during surgical operations. In particular, attention should be paid to the following.

1. Hematoma formation.
2. Localized skin necrosis due to tension.
3. Infection.
4. Keloid formation.
5. Poor cosmetic result.
6. Tissue reaction with suture.

Once applied, absorbable suture materials are digested by tissues during the healing process, while nonabsorbable suture materials resist enzymatic digestion and are normally removed after a period of time post surgery. For example, sutures used on face and head are usually removed after 5 days, those used on legs and abdomen are removed after 7–10 days, and for surgery on the back sutures are usually removed after 10–14 days.

During application, a specific knotting technique is required to handle and tie the sutures. With multifilament sutures, the nature of the material and the braided or twisted construction provide a high coefficient of friction and the knots remain as they are laid down. In monofilament sutures the coefficient of friction is relatively low, resulting in a greater tendency for the knot to loosen after it has been tied. In addition, monofilament synthetic polymeric materials possess the property of memory, with a tendency not to lie flat but to return to a given shape set by the material's extrusion process or the suture's packaging.

During surgery, suture knots must be properly placed in order to be secure. Speed in tying knots may result in less than perfect placement of the strands. In addition to variables inherent in the suture materials, considerable variation can be found between knots tied by different surgeons and even between knots tied by the same individual on different occasions. Complications associated with knots are well known. Problems with knots that have been reported in the literature include knot breakage and slippage, suture extrusion or "spitting," and infection, especially with braided sutures. Tightly approximated wounds and overly taut sutures have also been shown to cause ischemia and scarring, reduced wound strength and inflammation, and even dehiscence.

9.8 Test methods

Tensile strength, stiffness, easy handling properties, and good knotting security are the most important properties of surgical sutures. To assess these properties, tests can be carried out according to ASTM D3217-07 standards, which cover the measurement of the breaking tenacity of manufactured textile fibers taken from filament yarns, staple, or tow fibers, either crimped or uncrimped, and tested in either a double loop or as a strand formed into a single overhand knot. The breaking tenacity of conditioned and wet (immersed) fibers in loop and knot form is also included. When testing surgical sutures, the samples are first cut to a length of 6 cm. Both knotted and knotless samples should be prepared for each suture type. The samples are then separately put into tubes containing seven different medical liquids. They are kept at temperatures of 21 and 4 °C, because medical operations are often carried out at 21 °C and medical liquids are prepared at 4 °C, according to the Biomedical Laboratory of Stuttgart Textile Fiber and Research Institute.

Tensile strengths, tensile strains, and weight losses of the samples are then tested after an interval of 2 days. Different types of suture materials have considerably different performance profiles. For example, the strength of catgut sutures is strongly affected by temperature increase. Tensile strengths of catgut sutures are higher at lower temperatures. Catgut samples are also seriously deformed by blood.

Since suture materials are always exposed to blood when they are implanted into human body, during the tests they are subjected to medical liquids such as isotonic and dextrose solutions. It is also important to test how an implanted material will behave with or without knotting. Experimental results have shown that tensile strengths of both synthetic and natural sutures decrease in different solutions at different time intervals. Tensile strengths of samples at 21 °C are usually lower than those at 4 °C, with the temperature having a significant effect on tensile strength of sutures. In addition, tensile strengths of knotted specimens are usually lower than knotless specimens.

9.9 Summary

Surgical sutures are a type of commonly used medical textile product with a history dating back to ancient times. During the development of suture materials, it is important to engineer the fiber-forming polymers, the filament fibers, and the sutures to ensure critical requirements are met concerning knot safety, stretch capacity, tissue reactivity, and wound safety. Along with good clinical practice, high-quality surgical sutures can help to avoid clinical complications such as hematoma formation, localized skin necrosis due to tension, infection, keloid formation, poor cosmetic results, and tissue reaction with sutures.

Sources for further information

[1] Majno G. The healing hand: man and wound in the ancient world. Cambridge: Harvard University Press; 1991.
[2] Leaper DJ. Wound closure: basic techniques. EWMA J 2001;1(2):19–24.
[3] Tiyek I, Gemci R, Turkoglu S. Comparison of physical properties of sutures in medical liquids. Int J Phys Sci 2011;6(8):2158–68.
[4] Gemci R, Ulcay Y. Types and properties of surgical threads and strength differences between chrome and normal catgut. J Fac Eng and Archit Uludag University 2004;9(2):95–105.
[5] Rosin E, Robinson GM. Knot security of suture materials. Vet Surg 1989;18(4):269–73.
[6] Leaper DJ, Harding KG, editors. Wounds: biology & management. Oxford: Oxford Medical Publications; 1998.
[7] Capperauld I. Suture materials: a review. Clin Mat 1989;4:3–12.
[8] Leaper DJ, Pollock AV, Evans M. Abdominal wound closure: a trial of nylon, polyglycolic acid and steel sutures. Br J Surg 1977;64:603–6.
[9] Bucknall TE, Ellis H. Abdominal wound closure: comparison of monofilament nylon and polyglycolic acid. Surgery 1981;6:672–7.
[10] ASTM D3217–07: Standard Test Methods for Breaking Tenacity of Manufactured Textile Fibers in Loop or Knot Configurations.

Textiles for implants and regenerative medicine

10

10.1 Introduction

Textile materials have long been used for medical purposes—for example, as described in Chapter 9, sutures have a history dating back to ancient Egypt around 3000 BC. However, due to poor understanding in material and medical sciences, the early uses of fibers and textile materials in medical applications were mainly concerned with wound care, such as wound dressings. Because of the difficulties involved with surgery, it was not widely known that textile structures could be used as permanent implants. In recent years, with the advancement of medical science, fiber and textile materials are increasingly used as artificial organs for implantation into the body and for regenerative medicine, which is beginning to revolutionize many traditional therapies. Figure 10.1 shows some examples of implantable textile materials.

The human body is full of fibrous materials, such as collagen within the connective tissue, muscles, tendons, and nerves, which resemble the flexible nature of textile materials. When a surgical implant is required, textile-based materials have the combined properties of high strength and high flexibility, making them one of the best materials for constructing many types of artificial organs, with unique properties that cannot be obtained from metals, ceramics, or other forms of biomedical materials. Table 10.1 shows some examples of textile materials and structures used in different types of medical implants.

10.2 Characteristics of textile-based implant materials

A medical implant is a synthetic device, in contrast to a transplant, which is a transplanted biomedical tissue. Technically, an implant is a medical device manufactured to replace a missing biological structure, support a damaged biological structure, or enhance an existing biological structure. The surface of implants that contact the body might be made of a biomedical material such as titanium, silicone, or apatite depending on the functional requirement. Some implants are bioactive, such as subcutaneous drug-delivery devices in the form of implantable pills or drug-eluting stents. Other medical implants may contain electronic devices, such as artificial pacemakers and cochlear implants.

When used inside the body, the implant materials should not initiate undesired host response, nor cause any undesired reaction from neighboring or distant tissues. In practice, since a medical implant is an invasive material, its interaction with the surrounding tissue can lead to complications during or after surgery, such as infection,

Medical Textile Materials. http://dx.doi.org/10.1016/B978-0-08-100618-4.00010-8

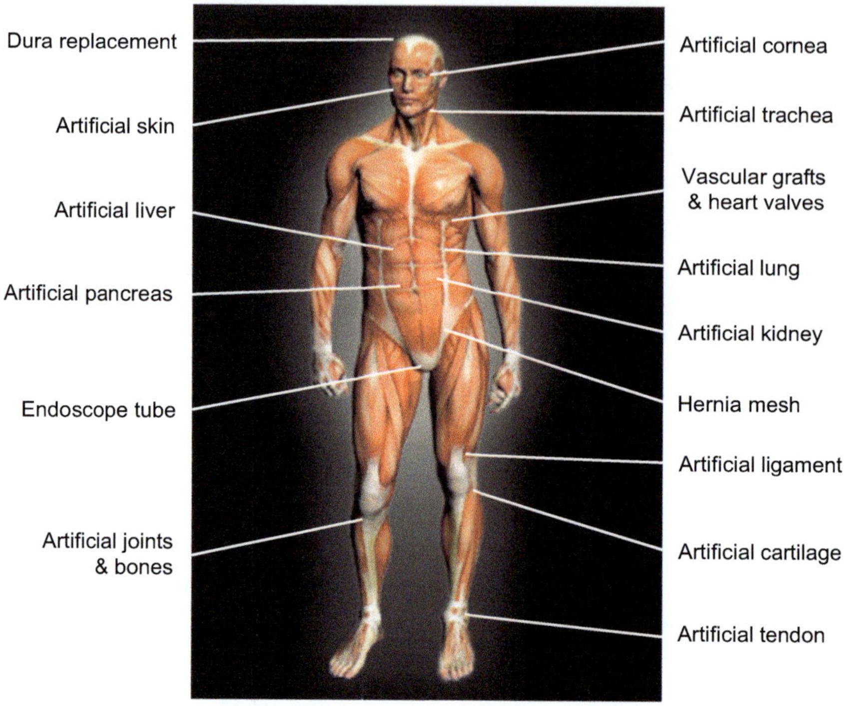

Figure 10.1 Some examples of implantable textile materials.

Table 10.1 Applications of textile materials and structures in medical implants

Applications	Materials	Textile structures
Artificial tendons	Polyester filaments	Braided, woven
Arteries	Polyester and Teflon filaments	Knitted, woven, nonwoven
Heart valves	Polyester filaments	Knitted, woven
Vascular grafts	Polyester and PTFE filaments	Warp knitted
Sutures	Polyester, nylon, silk, collagen filaments	Braided
Artificial bone	Carbon, polyacetal, polyethylene filaments	Braided, woven
Artificial ligaments	Polyester, silk, PTFE, polyethylene filaments	Braided, woven

inflammation, and pain. Implant-induced coagulation and allergic foreign body response may lead to rejection by the host body. The coagulation process is triggered by proteins that attach to the implant surface and lose their shapes, which may then trigger an immune system response when the body attempts to attack the implant in order to remove the foreign material. This can be accompanied by inflammation, sometimes leading to chronic inflammation where the implant is rejected and has to be removed from the body. Inflammation may occur as part of an immune response following the placement of surgical implants into the body, which results in the rapid dilation of local capillaries. As a result, the inflow of blood can swell the tissue and cause cell death. The excess blood, or edema, can also activate pain receptors at the tissue surrounding the medical implant.

10.3 Advantages of textile implant materials

Textile materials have been made into meshes for hernia repair and vascular prostheses for implantation, as well as cartilage, nerve, and liver regeneration. Compared to metals and ceramics, the main advantage of textile materials are their mechanical properties, flexibility, and elasticity, as well as the large varieties of possible designs and structures. For example, the porosity of textile materials can be adjusted in a wide range by varying the fiber diameter, yarn thickness, and fabric density, with the fiber diameter ranging from 1 μm up to more than 100 μm, and with different cross-sectional shapes such as round, trilobal, and star-shaped. In addition, the development of fiber science has generated endless possibilities for the chemical and physical modification of textile fibers and related products. For example, on the molecular level, polymers with a specified degradation profile can be used to make fibers that can degrade within a specified period of time. On the textile level, novel specialty fibers can be processed into woven, knitted, braided, and other forms of functional textile structures to meet specific end-use requirements. In this respect, various textile technologies have been used to engineer products with defined porosity, including pores between the yarn filaments and pores defined by the interloop space. Specifically, spacer fabrics have been used for scaffolding human and rodent cells like keratinocytes or hepatocytes, while embroidery technology can be used to control porosity and mechanical properties locally in a given textile geometry without any periodicity. Stitching techniques can be used for local drug delivery from a textile implant and for the development of composite textile materials with locally different degradation rates. A knitted structure might be the first choice where high elasticity is required, while woven fabric can be used for its dimensional stability.

Braided textile fabrics are a particularly useful form of implant materials. The integral structures of braided textiles enable them to endure twisting, shearing, and impact better than woven or knitted fabrics. Although braided fabrics exhibit poor stability under an axial compression, their high impact resistance and good stability under tension mean they can be designed for multidirectional conformity. Plain tubular braids may be used as prostheses for the replacement of injured ligaments in joints, such as the human knee.

Biologically, textile structures can be used to engineer biodegradable tissue scaffolds, where cellular components can generate new tissue through the production of extracellular matrix while the scaffold material provides structural integrity and mechanical stability. The textile materials can be treated in a number of ways to obtain the required structure and porosity, which are the key elements governing the formation of new tissue and subsequent neovascularization in vivo.

10.4 Artificial ligaments and tendons

One of the most important human ligaments is the anterior cruciate ligament (ACL), which is a dense, highly organized, cable-like tissue composed of collagens, elastin, proteoglycans, water, and cells. The human ACL has an average length of 27–32 mm and a cross-sectional area of 44.4–57.5 mm^2. It contains a hierarchical structure with different levels of organization, including collagen molecules, fibrils, fibril bundles, and fascicles that run parallel to the long axis of the tissue. The ACL is the major intraarticular ligament of the knee and is critical to the normal kinematics and stability of the body. It controls motion by connecting the femur to the tibia and stabilizing the joint, preventing abnormal types of motion, and in general supporting and strengthening the knee.

The ACL is the most commonly injured ligament. After injury, ligament healing can be divided into three phases: inflammation, cellular proliferation and matrix repair, and remodeling. When injury causes a full rupture of the ligament or its detachment, insertion-point surgery is required. If the ligament is not repaired, the loading of the joint leads to abnormal stress on the articular cartilage, which can lead to early osteoarthritis.

Arthroscopically assisted ACL reconstruction has been widely used for patients with anterior knee laxity, and good clinical results have been obtained following advances in arthroscopic surgery. The bone-patellar tendon-bone (BPTB) autograft has become an established procedure in ACL reconstruction, while a hamstring tendon graft has also been used as an alternative to the BPTB.

The use of synthetic material for ligament reconstruction was developed in the 1980s. However, after a preliminary period of enthusiasm for these implants, their popularity declined because of the high device failure rate and reactive synovitis caused by wear particles. Following many years of new development, the Ligament Advanced Reinforcement System artificial ligament has recently been reported to be a suitable device due to its special design, and satisfactory clinical results have been obtained following its use in ACL reconstruction.

It should be pointed out that during the development of artificial ligament knowledge about ligament structure and its components is extremely important, because it is the interaction between these components and their arrangement in the tissue that give ligaments their unique mechanical properties. Ligaments display triphasic behavior when exposed to strain. First, in the nonlinear or toe region the ligament exhibits a low amount of stress per unit strain. This is followed by an area noted for its increase in stress per unit strain, called the linear region. The last region is the yield and failure region, which shows a slight decrease in stress per unit strain and marks the failure of the ligament. This unique behavior is due to the components of the ligament and their arrangement in the tissue. When force is first applied to the tissue it is transferred to the collagen fibrils, which results in the lateral contraction of the fibrils, the release of water, and straightening of the crimp pattern in the collagen fibrils. Once the crimp pattern is straightened, the force is applied directly to the collagen molecules when the collagen triple helix is stretched and interfibrillar slippage occurs, which results in an increase in stress per unit strain. Finally, the collagen fibers in the ligament fail by defibrillation, causing a decrease in stress per unit strain and tissue failure.

10.5 Surgical meshes for hernia repair

Historically, surgical openings were often closed by suturing, which is still commonly used in many surgical treatments. A "tension-free repair" with meshes is a relatively new therapy that can allow faster recovery, using a mesh to reinforce the body tissue. When surgical meshes were first implanted in the 1950s they were made of polyethylene terephthalate (PET) multifilaments, but in the 1960s polypropylene monofilaments became available as mesh materials, and have since become the most appropriate choice for surgical meshes, partly because PET loses strength over time (>10 years). However, as strength may not be needed over such a long period, PET meshes are still available today. Most of the meshes are warp-knitted fabrics, which have some elasticity and can allow the meshes to be cut into the desired size.

In the surgical practice, hernia meshes are the most frequently used textile implants worldwide, with more than a million implanted every year. Hernias occur quite often, mainly in the form of organs and tissues penetrating through the abdominal wall. These protrusions are not only painful but can also damage the penetrating organ, such as the gut, and therefore closure of the defect is important. It is one of the most common surgical procedures, with more than 1 million operations performed in the US and over 250,000 in Germany each year.

Today, meshes for hernia repair are well-established medical devices that are widely used around the world. In order to overcome chronic pain and formation of scar tissue, which are encountered during hernia implantation, new materials and devices have been developed in this field, for example, by using degradable or partially degradable meshes made of polyglycolic acid (PGA) or a mixture of 50% PGA and 50% polypropylene.

10.6 Vascular prostheses

Artificial blood vessels are another important type of medical textile materials, and are mainly inserted into an existing artery. As is commonly known, arteriosclerosis is the result of degenerative changes of the vessel wall, which can lead to thickening, loss of elasticily, and finally stenosis or even occlusion of the blood vessel. The stenosis of coronary arteries is a frequently encountered medical problem, and is often treated by using natural vessels (mainly veins from the leg) to make a bypass.

Historically, the idea of using textile vascular prostheses came up in the US in 1952, when their suitability was demonstrated in experiments where the abdominal aneurysm was successfully replaced in 17 patients with textile structures sutured to tubular forms. After many experiments, it has been found that PVC is not suitable for long-term implants since the plasticizer used to make the fiber flexible is toxic and can leach out, while polyamide is also unsuitable because it degrades over time and can cause leakage. The most commonly used material for making vascular prostheses is PET, thanks to the famous cardiac surgeon Michael DeBakey, who first introduced PET for this application. It has turned out to be very blood compatible and biocompatible, and is therefore still widely used today. Since the first PET fibers were made by DuPont under the trade name Dacron®, today these arterial prostheses are still called "Dacron prostheses" by many surgeons.

During the production process, the prostheses are processed from multifilament yarns by warp knitting or weaving. Normal weft knitting is no longer used because of the risk of the knitting unraveling. While the first prostheses were woven before being sutured to tubes, today special machines based on narrow fabric looms can allow the production of tubular fabrics. Woven prostheses are very dense, and have low porosity and elasticity but a strong breaking strength, making them suitable for use in high-pressure areas like the aortic arc or in patients with a disturbed clotting system. It is important that the ends of the woven tube are fixed in order to avoid the dissolution of the textile structure at the anastomosis where the prosthesis is sutured to the native blood vessel.

Vascular prostheses fabricated by warp knitting are highly recommended because the radial elastic vessel walls actively support the blood transport of the heart by the so-called Windkessel effect. Tubular structures can be directly produced on double-needle-bar Raschel machines, even as bifurcation. The finishing agents used during fiber spinning, knitting, or weaving must be thoroughly removed by solvent extraction before the prostheses are sterilized by appropriate methods. The porosity of the warp-knitted structure is reduced with boiling water by using highly shrinkable yarns. Lastly, the radial stability of the prostheses is fixed by pleating it on a waved rod under heat, giving the final shape of the device.

It is known that small-diameter coronary prostheses with diameters <6 mm are not effective, since the body's reaction to foreign material will cause an occlusion of the device in a very short time. More commonly, textile-based prostheses are used where diameters of more than 6 mm are to be replaced. The most common textile prosthesis is used for the abdominal aortic aneurysm (AAA), a bulge of the aorta close to the bifurcation into the two iliac arteries leading into the legs. AAA is caused by a dilatation of the weakened vessel wall and occurs most commonly in older men (>60 years old). The disease is normally symptom free, but may end in a rupture of the aorta which leads to death within minutes due to internal bleeding. It is estimated that worldwide 4.5 million people are living with an AAA, and in the USA about 15,000 people die every year from AAA rupture.

During the application of vascular prosthesis, the porosity of the textile structure would cause bleeding if the pores are not temporarily closed. Pore closure used to be achieved by soaking the artificial vessel in blood and allowing it to clot. Today, most prostheses are coated with a gel, usually gelatin. In both cases the closing material is degraded by the body over time and replaced by proteins deposited from the blood, which is called neointima or pseudointima.

Vascular prostheses have demonstrated a good compliance over the last 50 years, but the operation procedure involves some risk of failure and death of the patient. This increases with the length of the aneurysm, the morbidity of the original vessel, and the inclusion of the renal arteries in the aneurysm. To overcome these problems, alternative materials have been developed to engineer vascular prostheses, the most promising being stent grafts developed in the mid- to late 1990s. The typical stent is a metallic tubular grid or braid made of steel or nitinol, which is a shape-memory alloy of titanium and nickel. The stent is generally expanded into blocked coronary arteries to keep them open. If the open structure of a stent is coated, it is called a stent graft. In an AAA, stent grafts were developed to avoid opening up the abdomen—a stent graft can be introduced percutaneously by minimal invasive techniques through the femoral arteries by using catheters, a process called endovascular aneurysm repair.

To improve their blood compatibility, the inner surfaces of prostheses are seeded with endothelial cells to reconstitute the natural inner surface structure of blood vessels. These cells prevent blood coagulation and the formation of a thrombus, but in practice it is difficult to hold the cells on the artificial surface for long periods. Rather than preseeding the prosthesis, a new method has been developed to attract endothelial cells from the blood stream on to the inner wall of the prosthesis. New processes have also been developed to suppress bacterial growth, because intraoperative infections are still a general problem causing persistent biofilms on implants.

A major problem of arterial grafts is the induction of clotting by the graft's surface. To circumvent this, grafts are usually clotted with the patient's own blood before implantation to reduce seepage through the knitted structure. Preclotting has been found useful in delaying or preventing thrombosis. Albumin has been cross-linked around the graft and deposited by soaking the graft. Heparin has also been used as a cationic surfactant to improve blood compatibility.

10.7 Textiles for regenerative medicine

Regenerative medicine is concerned with the process of replacing, engineering, or regenerating human cells, tissues, or organs to restore or establish normal function. It stimulates the body's own repair mechanisms to heal previously irreparable tissues or organs and restore functionality, and is one of the most promising research areas in medical sciences.

There are two basic approaches to restore the functions of damaged tissues and organs in the human body. First, damaged tissues can be replaced by biological substitutes that are grown outside the body through tissue engineering, and textile materials are ideally suited to provide the scaffold for cell growth. Second, textile materials can be used to carry biomaterials, cells, and signaling molecules to guide regeneration inside the body rather than building the whole tissue in a laboratory. In the last few decades tissue engineering has been applied to nearly every organ or tissue, and the complex regulation of living tissue is fairly well understood. A major problem remains the integration and connection of "engineered tissue" into the existing environment in the body. In contrast, regeneration of tissue or organs can proceed more smoothly inside the body through the stimulation of repair mechanisms by various types of mediators. For example, stem cells found in a newborn child's umbilical cord blood hold great promise in cardiovascular repair. In animal models of myocardial infarction, cord-blood stem cells have shown the ability to migrate selectively to injured cardiac tissue, improve vascular function and blood flow at the site of injury, and improve overall heart function.

Biologically grown tissues are often engineered by implanting or seeding cells into an artificial structure capable of supporting three-dimensional tissue formation. These so-called scaffolds are critical to replicating the in vivo milieu and allowing cells to influence their own microenvironments. Scaffolds usually serve at least one of the following purposes.

1. Allow cell attachment and migration.
2. Deliver and retain cells and biochemical factors.
3. Enable diffusion of vital cell nutrients and expressed products.
4. Exert certain mechanical and biological influences to modify the behavior of the cell phase.

The surface structure of the scaffold material is important for tissue engineering as well as in regenerative medicine, since, as living organisms, cells interact closely with the surface they adhere to. When cells bind to the scaffold, they do so by first depositing matrix proteins on the material surface, and these proteins can be modified by the chemical nature of the surface. To improve cell adherence, the scaffold can be coated with hydrophilic components or covalently bond with signaling molecules. It should be noted in this context that tissues are very dynamic and a hydrophilic surface allows a reversible attachment of matrices and cells, thus supporting normal tissue formation. For this reason, more and more natural materials are being considered for tissue engineering, such as protein-based (collagen, fibrin) and carbohydrate-based (alginate, chitosan, etc.) fibers. The choice is mainly dependent on the intended tissue or organ.

During the preparation of artificial tissue, specific cells are first isolated from the patient's body and are then proliferated in the laboratory and seeded into the scaffolds, before being brought back into the injured tissue. In practice, the creation of functional tissues and biological structures in vitro requires extensive culturing to promote survival, growth, and inducement of functionality. The basic requirements of cells, which include oxygen, pH, humidity, temperature, nutrients, and osmotic pressure, must be maintained in culture. In standard cell culture, diffusion is often the sole means of nutrient and metabolite transport. However, as a culture becomes larger and more complex, as is the case with engineered organs and whole tissues, other mechanisms must be employed to maintain the culture, such as the creation of capillary networks within the tissue. Another issue with tissue culture is introducing the proper factors or stimuli required to induce functionality. In many cases, simple maintenance culture is not sufficient. Growth factors, hormones, specific metabolites or nutrients, and chemical and physical stimuli are sometimes required. For example, certain cells respond to changes in oxygen tension as part of their normal development, such as chondrocytes, which must adapt to low-oxygen conditions or hypoxia during skeletal development. Others, such as endothelial cells, respond to shear stress from fluid flow, which is encountered in blood vessels. Mechanical stimuli, such as pressure pulses, seem to be beneficial to all kind of cardiovascular tissue, including heart valves, blood vessels, and pericardium.

In tissue or organ repair with regeneration therapy, a crucial subject is angiogenesis, because newly formed tissues have to be connected to the blood stream. fibroblast growth factor (FGF) and vascular endothelial growth factor are important in the stimulation of vessel growth. Some gases also have signaling functions, such as nitric oxide and carbon monoxide. It is a major challenge of regenerative medicine to include these signaling molecules in devices and deliver them at the right moment and in the intended location.

There have been many successful applications of regenerative therapy. A typical example is the use of textile-based materials in repairing damaged nerves. Following a trauma, peripheral nerves often fail to regenerate the lost gaps and a cut nerve cannot reconnect. Although this problem can be solved by autologous nerve grafts, a promising alternative is to use nerve guide tubes to restore the lost gaps. The creation of artificial conduits is also known as entubulation, because the nerve ends and intervening gap are enclosed within a tube composed of biological or synthetic materials. Whether the conduit is in the form of a biologic tube, synthetic tube, or tissue-engineered conduit, it should facilitate neurotropic and neurotrophic communication between the proximal and distal ends of the nerve gap, block external inhibitory factors, and provide a physical guidance for axonal regrowth. Many commercial products have already been developed and used in clinical applications. For example, NeuraGen® is a collagen-based tube while Neurotube® is a woven PGA tube. A recent development used a block-copolymer made from trimethylene carbonate and caprolactone to engineer hollow-fiber membranes with a diameter of 1 mm that can be used as nerve guide tubes. These tubular resorbable nerve guides were seeded with Schwann cells, and were able to regenerate nerves even across a large gap of 2 cm in vivo.

10.8 Regulatory aspects

Because of their direct and sustained contact with the human body, medical implant materials can generate a large number of safety risks either through their close contact with the body or from the release of materials involved in their biodegradation. To ensure safety and functional performances, textile materials for medical implants and regenerative therapies are extensively tested in vitro and in vivo. In vitro testing of textile implants includes a broad range of chemical, mechanical, and biological tests, and standards have been developed for most of these. ISO 10993, titled "Biological Evaluation of Medical Devices," summarizes the in vitro test methods for medical devices such as textile implants. There are strict regulations governing the clinical use of medical implants and regenerative therapies. Depending on the nature of the treatment, these products may be classified as a medical device, such as wound-dressing material that supports the regeneration of skin, a pharmaceutical, such as a bone morphogenetic protein, a cell therapy, such as chondrocyte transplanted into the knee, or a combination product.

In the United States the Center for Biologics Evaluation and Research is responsible for regulating human cells, tissues, and cellular and tissue-based products. Cells in scaffolds are classified as combination products, and may be classed as either biologics or medical devices. Because of their complex nature, it has been difficult for national regulatory agencies to devise regulations for these new types of therapies. In 2007 therapies based on genes or living cells or tissues were classified in Europe as "Advanced Therapy Medicinal Products," which means that these products will be handled like pharmaceuticals and approved by the European Medicines Agency. This results in strong regulation not only on the production side but also at the location of application, for example, the clinic.

10.9 Summary

Textile materials are soft, flexible, and strong, and thus resemble the physical and mechanical properties of many biological structures within the human body. By careful control of the polymer and fiber structures and appropriate surface treatment, textile-based materials can be engineered into two- or three-dimensional structures that can be used as artificial ligaments, hernia meshes, vascular prostheses, and many other novel implants. In vitro and in vivo studies have demonstrated the effectiveness of these novel biomaterials. In addition to being implanted into the human body as passive materials, textile-based materials can be engineered to carry living cells and other bioactive agents that can stimulate the body's biological functions to regenerate tissues and organs. This unique interaction between implanted materials and the surrounding tissues offers a fascinating area of research and development for novel materials that can help restore and enhance human health.

Sources for further information

[1] Bartels VT, editor. Handbook of Medical Textiles. Cambridge: Woodhead Publishing Ltd; 2011.

[2] Lyons I. Biomedical science: lecture notes. Oxford: Wiley Blackwell; 2011.

[3] Laurencin CT, Freeman JW. Ligament tissue engineering: an evolutionary materials science approach. Biomaterials 2005;26:7530–6.

[4] Feller J, Webster K, Gavin B. Early post-operative morbidity following anterior cruciate ligament reconstruction: patellar tendon versus hamstring graft. Knee Surg Sports Traumatol Arthrosc 2001;9:260–6.

[5] Chastan P. Tension free open inguinal hernia repair using an innovative self gripping semi-resorbable mesh. J Minim Access Surg 2006;3:139–43.

[6] DeBakey ME. Clinical application of a new flexible knitted dacron arterial substitute. Am Surg 1958;24(12):862–9.

[7] Dieval F, Mathieu D, Durand B. Influence of textile structure on longitudinal ruptures' localization of the vascular prostheses. Text Res J 2008;78(5):427–38.

[8] Greenhalgh RM, Powell JT. Endovascular repair of abdominal aortic aneurysm. N Engl J Med 2008;358(5):494–501.

[9] Hollinsky C. Comparison of a new self-gripping mesh with other fixation methods for laparoscopic hernia repair in a rat model. J Am Coll Surg 2009;6:1107–14.
[10] Langer R, Vacanti JP. Tissue engineering. Science 1993;260(5110):920–6.
[11] Mason C, Manzotti E. Regenerative medicine cell therapies: numbers of units manufactured and patients treated between 1988 and 2010. Regen Med 2010;5(3):307–13.
[12] Moghe AK, Gupta BS. Small diameter blood vessels by weaving: prototyping and modeling. J Text Inst 2008;99(5):467–77.
[13] Harris DT. The potential of cord blood stem cells for use in regenerative medicine. Expert Opin Biol Ther 2007;7(9):1311–22.
[14] Upchurch GR, Schaub TA. Abdominal aortic aneurysm. Am Fam Physician 2006; 73(7):1198–204.
[15] Phillips JB, Bunting SC, Hall SM, Brown RA. Neural tissue engineering: a self-organizing collagen guidance conduit. Tissue Eng 2005;11(9–10):1611–7.
[16] Schmidt CE, Leach JB. Neural tissue engineering: strategies for repair and regeneration. Annu Rev Biomed Eng 2003;5:293–347.

Antimicrobial dressings for the management of wound infection

11

11.1 Introduction

Since wounds are broken skin with the loss of normal skin functions, the invasion of bacteria becomes a serious problem in wound healing. Wounds exude a large amount of fluid, which is warm and contains a variety of nutritional components that can benefit bacterial growth. Wound infection and cross-infection commonly occur in hospital wards. These infections complicate patient illness, cause anxiety, increase patient discomfort, and in the worst cases can lead to death of the patient.

Depending on the level of bacteria colonization, wound infection can be classified into the following stages.

1. Wound contamination: the presence of bacteria within a wound without any host reaction.
2. Wound colonization: the presence of bacteria within the wound which do multiply or initiate a host reaction.
3. Critical colonization: multiplication of bacteria causing a delay in wound healing, usually associated with an exacerbation of pain but with no overt host reaction.
4. Wound infection: the deposition and multiplication of bacteria in tissue with an associated host reaction.

Among the various types of bacteria that can cause wound infection, *Staphylococcus aureus* is a gram-positive bacterium that exists as a skin commensal and has the ability to cause a wide range of infections, from localized skin eruptions to life-threatening conditions such as bacteremia, endocarditis, and pneumonia. *Staphylococcus aureus* is one of the most common causes of hospital-acquired infections, and its pathogenicity is due to its production of coagulase, an enzyme that clots plasma and thus inhibits host defense mechanisms. In recent years methicillin-resistant *S. aureus* (MRSA) has attracted much attention in the medical field because of its deadly nature. Since it was first reported in the UK in the 1980s, many different strains of MRSA have been found, affecting a large number of individuals in many different healthcare settings. The degree to which people are affected ranges in severity from simple wound colonization, which does not need to be treated aggressively, to systemic infection, such as bronchopneumonia, which may be fatal.

The cost of managing the problems associated with infection is considerable. According to a report by the UK's National Audit Office, at any given time about 9% of hospital patients have a nosocomial infection, costing the National Health Service as much as £1 billion per annum and contributing to the death of an estimated 5000 people every year. The US Office of Technology Assessment reported that antibiotic-resistant infections caused by six species of bacteria in the US hospitals cost the country at least $1.3 billion a year.

Medical Textile Materials. http://dx.doi.org/10.1016/B978-0-08-100618-4.00011-X

The report of the UK working party on hospital infection, and more recently the report by the National Audit Office, recommended that despite practical problems, where infection control facilities may be inadequate or in situations where MRSA has become endemic, active intervention to prevent the further spread of the organism is of benefit and should be encouraged. Wound dressings that readily permit strikethrough or shed fibers on removal should be avoided, as they may transmit contaminated particles that could easily be carried around the room under air currents, contaminating the adjacent surfaces. The use of irrigation solutions to remove adherent dressings may also increase the potential for infected material to be transferred from the wound to the surrounding area, either in droplets that bounce off the wound surface when a jet of solution is applied with force by means of a syringe or in a gentle trickle that runs down the patient's leg. Semipermeable dressings such as films, film–foam combinations, and hydrocolloids, which effectively seal off the peri-wound area, may help to prevent the passage of contaminating organisms both into and out of a wound.

The most effective way to control the spread of bacteria from wound sites is to incorporate antimicrobial agents into wound dressings. Over the years many antimicrobial materials have been used in wound management, including chlorhexidine, honey, hydrogen peroxide, iodine, proflavine, silver, and many other novel materials.

11.2 Topical antimicrobial agents in wound care

For patients with infected wounds, two basic treatment protocols are available for medical practioners: systemic administration of antibiotics, and topical application of antimicrobial agents on the wounded area directly. While systemic antibiotics are outside the scope of this chapter, over the years many topical antimicrobial agents have been used to treat wound infection. These are briefly summarized below. In addition to these main materials, other antimicrobial agents that can be used in wound management include tea tree oil and gentian violet ointment. Extracts of tea are also said to have inhibitory effects on MRSA.

11.2.1 Chlorhexidine

Chlorhexidine is available as diacetate, digluconate, and dihydrochloride, with digluconate most frequently used in wound management. Chlorhexidine was discovered in 1946 and introduced into clinical practise in 1954. It has rapid bactericidal activity against a wide spectrum of bacteria, damaging outer cell layers and the semipermeable cytoplasmic membrane to allow leakage of cellular components. It also causes coagulation of intracellular constituents, depending on concentration. Chlorhexidine is widely used as an antiseptic in hand washing and as a surgical scrub, but in wound management its application has been limited largely to irrigation.

11.2.2 Hydrogen peroxide

Hydrogen peroxide has been widely used as an antiseptic and disinfectant. A 3% solution has most often been used to clean wounds. It is a clear, colorless liquid that decomposes when in contact with organic matter. It has a broad spectrum of activity against bacteria, with a greater effect on gram-positive species than gram negatives. Hydrogen peroxide functions as an oxidizing agent by producing free radicals that react with lipids, proteins, and nucleic acids to affect cellular constituents.

11.2.3 Iodine

Iodine is an element that was discovered in 1811. It is a dark-violet solid that dissolves in alcohol and potassium iodide. Its first reported use in treating wounds was by Davies in 1839, and later it was used in the American Civil War. Early products caused pain, irritation, and skin discoloration, but since 1949 the development of iodine compounds such as povidone iodine and cadexomer iodine has yielded safer and less painful formulations.

Povidone iodine is a polyvinylpyrrolidone surfactant/iodine complex. Cadexomer iodine is composed of beads of dextrin and epichlorohydrin that carry iodine. Both release sustained low concentrations of free iodine whose exact mode of action is not known, but involves multiple cellular effects by binding to proteins, nucleotides, and fatty acids. Iodine is thought to affect the protein structure by oxidizing the S—H bonds of cysteine and methionine, reacting with the phenolic groups of tyrosine and with N—H groups in amino acids, such as arginine, histidine, and lysine, to block hydrogen bonding. It reacts with bases of nucleotides to prevent hydrogen bonding and alters membrane structure by reacting with C=C bonds in fatty acids. It has a broad spectrum of activity against bacteria, mycobacteria, fungi, protozoa, and viruses.

11.2.4 Proflavine

Proflavine is a bright-colored acridine derivative that was extensively used during the Second World War in the treatment of wounds. Modern use is as a prophylactic agent in surgical wounds packed with gauze soaked in proflavine hemisulfate solution. It is an intercalating agent that inhibits bacteria by binding to DNA and prevents unwinding prior to DNA synthesis.

11.2.5 Honey

Honey has been used in wound-management practice for a long time and many therapeutic properties have been attributed to it, including antibacterial activity and the ability to promote healing. Evidence of antibacterial activity is extensive, with more than 70 microbial species reported to be susceptible.

The use of honey for treating infected wounds is increasing in popularity and a number of dressings or preparations containing it are now available, some of which have been shown to possess good antimicrobial activity against a wide range of pathogenic organisms, including resistant strains.

11.2.6 Silver

Silver has a long history as an antimicrobial agent, especially in the treatment of burns. Metallic silver is relatively unreactive, but in the aqueous environment silver ions are released and their antimicrobial activity depends on the intracellular accumulation of low concentrations of the silver ions. These avidly bind to negatively charged components in proteins and nucleic acids, thereby effecting structural changes in bacterial cell walls, membranes, and nucleic acids that affect viability. In particular, silver ions are thought to interact with thiol groups, carboxylates, phosphates, hydroxyls, imidazoles, indoles, and amines either singly or in combination, so that multiple deleterious events rather than specific lesions simultaneously interfere with microbial processes. Hence silver ions that bind to DNA block transcription, and those that bind to cell-surface components interrupt bacterial respiration and adenosine triphosphate synthesis.

Silver sulfadiazine (SSD) is one of the most widely used silver compounds in antimicrobial formulations. When introduced in 1968, SSD was recommended as a topical treatment agent for the prevention of pseudomonad infections in burns. It has since been demonstrated to possess broad-spectrum antibacterial, antifungal, and antiviral activity.

11.3 Main types of antimicrobial wound dressings

In developing wound dressings with antimicrobial functions, it is important to note that in addition to these functions wound dressings have many other performance criteria, such as the need to protect the wound surface, the ability to absorb wound exudates, ease of application and removal, etc. Traditionally, wound dressings are made from textile materials, since they share the same functional requirement of being able to protect the underlying object. Over the years many methods have been used to combine antimicrobial agents with base materials to make wound dressings with antimicrobial properties. These methods are summarized below.

11.3.1 Antimicrobial creams and ointments

Antimicrobial agents can be combined with a suitable thickening and gelling agent to prepare creams and ointments with antimicrobial properties. For example, Betadine Solution and Betadine Cream contain 10 and 5% povidone-iodine respectively. It has been shown that these products are effective against MRSA.

In another example, antimicrobial silver ions are combined with a hydrogel. SilvaSorb Silver Antimicrobial Wound Gel is an amorphous gel wound dressing for use in moist wound healing. Its SilvaSorb MicroLattice® technology maintains an optimally moist wound environment by either absorbing wound drainage or donating moisture while it delivers antimicrobial ionic silver. SilvaSorb Silver Antimicrobial Wound Gel can act as an effective antimicrobial barrier.

11.3.2 Textile fabric soaked with antimicrobial agents

A simple way to prepare antimicrobial wound dressings is to soak a traditional dressing in a hydrogel-containing antimicrobial agents. For example, some paraffin gauze dressings are medicated with antibiotics or antimicrobial agents for the treatment or prevention of infection. Sofra-Tulle from Hoechst contains framycetin, while Bactigras from Smith & Nephew Medical and Serotulle from Seton Healthcare contain 0.5% chlorhexidine acetate. Inadine from Johnson & Johnson Medical, which is similar in appearance to the other medicated-tulle products, contains povidone-iodine in a polyethylene glycol base coated on a knitted viscose fabric. Urgotul SSD dressing comprises a polyester mesh impregnated with carboxymethylcellulose, vaseline, and 3.75% silver sulfadiazine.

11.3.3 Antimicrobial-coated textile substrate

Antimicrobial agents can be coated on to the surface of a textile substrate to produce antimicrobial wound dressings. For example, Acticoat consists of two layers of a silver-coated, high-density polyethylene mesh, enclosing a single layer of an apertured nonwoven rayon and polyester fabric. These three components are ultrasonically welded together to maintain the integrity of the dressing while in use. Silver is applied to the polyethylene mesh by a vapor-deposition process which results in the formation of microscopic crystals of metallic silver. Upon activation with water, Acticoat provides a rapid and sustained release of silver ions within the dressing and to the wound bed.

Silverlon is a knitted fabric dressing that has been silver-plated by a chemical plating technique through reduction and oxidation. This technique coats the entire surface of each individual fiber from which the dressing is made, resulting in a very large surface area for the release of ionic silver.

11.3.4 Textile fibers containing antimicrobial agents

Antimicrobial compounds can be mixed with the spinning solution during the spinning process to produce fibers with antimicrobial properties. For example, AlphaSan RC5000 is a silver–sodium–hydrogen–zirconium–phosphate compound with an average particle size of about one micron. It consists of a three-dimensional repeating framework of sodium hydrogen zirconium phosphate, with many equally spaced cavities containing silver. Silver (at 3.8% by weight) provides the main antimicrobial properties, while the framework matrix acts to distribute silver evenly without clumping or pooling throughout the individual fibers where the AlphaSan particles are added. When AlphaSan RC5000 is mixed with sodium-alginate solution, the fine particles can be evenly distributed in the spinning solution under a high rate of shearing. Because the particles are very fine, they can be suspended uniformly while the solution is extruded to form fibers. Since the sodium–hydrogen–zirconium–phosphate framework prevents the silver ions from oxidizing the alginate, this type of silver-containing alginate fiber remains white even after sterilization through irradiation. Figure 11.1 shows the photomicrograph of alginate fiber with the AlphaSan RC5000 particles uniformly distributed inside the fiber structure.

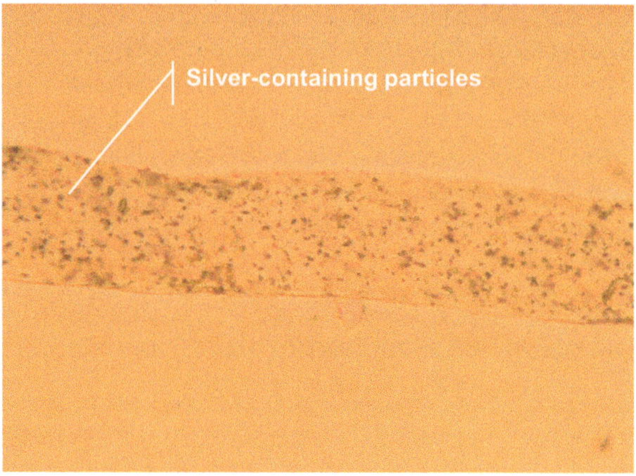

Figure 11.1 Photomicrograph of an alginate fiber showing the AlphaSan RC5000 particles uniformly distributed inside the fiber structure.
Source: Advanced Textiles for Woundcare (Chapter 7: Antimicrobial textiles in managing wound infection). Woodhead Publishing, 2009.

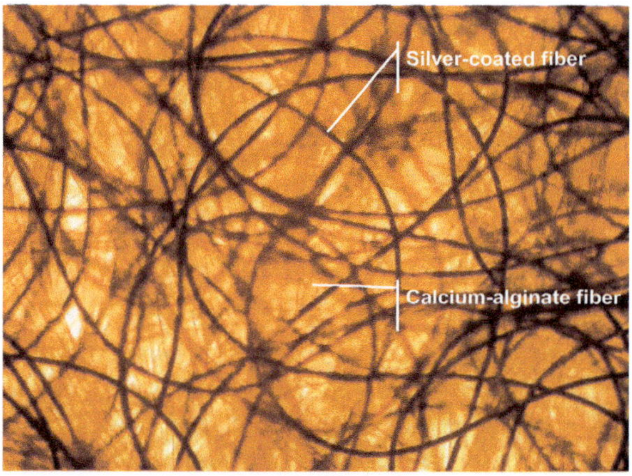

Figure 11.2 Photomicrograph of Silvercel wound dressing.
Source: Advanced Textiles for Woundcare (Chapter 7: Antimicrobial textiles in managing wound infection). Woodhead Publishing, 2009.

11.3.5 Textile composite containing antimicrobial fibers

Antimicrobial fibers can be mixed with traditional textile materials to produce antimicrobial wound dressings. For example, Silvercel from Johnson & Johnson combines the potent broad-spectrum antimicrobial action of a silver-coated nylon fiber with the enhanced exudate-management properties of alginate fibers. Because of the sustained release of silver ions, the dressing acts as an effective barrier and helps reduce infection. As shown in Figure 11.2, the antimicrobial properties are built in through the use of X-Static silver-coated fibers blended into the nonwoven structure.

The Silvercel dressing has been proven effective in vitro against 150 clinically isolated microorganisms, including antibiotic-resistant strains.

11.3.6 Other novel methods

Actisorb Silver 220 consists principally of activated carbon impregnated with metallic silver, produced by heating a specially treated fine viscose fabric under carefully controlled conditions. The carbonized fabric is enclosed in a sleeve of spun-bonded nonwoven nylon fabric, sealed along all four edges to facilitate handling and reduce particle and fiber loss. When applied to a wound, the dressing adsorbs toxins and wound-degradation products as well as volatile amines and fatty acids responsible for the production of wound odor. Bacteria present in an wound exudate are also attracted to the surface of the dressing, where they are killed by the antimicrobial activity of the silver.

Aquacel Silver consists of a fleece of sodium carboxymethylcellulose fibers containing 1.2% ionic silver. In the presence of exudate, the dressing absorbs liquid to form a gel and silver ions are released into the exudate, acting as the antimicrobial agent.

11.4 Silver-containing wound dressings

11.4.1 The antimicrobial properties of silver

Silver has broad-spectrum antimicrobial properties and a low level of toxicity to the human body. In recent years, as bacterial resistance to antibiotics became common and more and more attention was paid to cross-infection in hospital wards, wound dressings with antimicrobial properties became increasingly popular. Silver is gaining in importance as an effective antimicrobial component in advanced wound dressings.

The antimicrobial action of silver products has been directly related to the amount and rate of silver released and its ability to inactivate target bacterial and fungal cells. In various laboratory and clinical studies it has been found that metallic silver does not possess significant antimicrobial potency, while silver ions are highly antimicrobial. The oligodynamic microbicidal action of silver compounds at low concentrations probably does not reflect any remarkable effect of a comparatively small number of ions on the cell, but rather the ability of bacteria, trypanosomes, and yeasts to take up and concentrate silver from very dilute solutions. Therefore bacteria killed by silver may contain 10^5–10^7 Ag^+ per cell, the same order of magnitude as the estimated number of enzyme-protein molecules per cell.

Chemically, metallic silver is relatively inert, but its interaction with moisture on the skin surface and with wound fluids leads to the release of silver ions and their biocidal properties. Silver ions are highly reactive moieties and avidly bind to tissue proteins, causing structural changes in bacterial cell walls and intracellular membranes, which lead to cellular distortion and loss of viability. Silver ions bind to and denature bacterial DNA and RNA, thereby inhibiting replication.

A recent study demonstrated the inhibitory action of silver on two strains of gram-negative *Escherichia coli* and gram-positive *S. aureus*. It was found that exposure to silver nitrate led to the formation of electron-light regions in their cytoplasm and condensation of DNA molecules, with granules of silver observed in the cytoplasm. RNA and DNA damage and protein inactivation seemed to be the principal mechanisms for bacteriostasis. Silver-related degenerative changes in bacterial RNA and DNA, mitochondrial respiration, and cytosolic protein lead to cell death.

The action of silver ions on cell walls is illustrated by reference to the yeast *Candida albicans*. Silver has been shown to inhibit the enzyme phosphomannose isomerase by binding cysteine residues. This enzyme plays an essential role in the synthesis of the yeast cell wall, and its defects lead to the release of phosphate, glutamine, and other vital nutrients.

11.4.2 Types of silver compounds used in wound dressings

Silver is a group 11 element (formerly group Ib) of the periodic table and exists as two isotopes, ^{107}Ag and ^{109}Ag, in approximately equal proportions. In solution, silver exhibits three oxidation states, Ag^{+}, Ag^{++}, and Ag^{+++}, each capable of forming inorganic and organic compounds and chemical complexes. Compounds involving Ag^{++} or Ag^{+++} are unstable or insoluble in water.

The silver compounds used in wound dressings can be divided into three groups.

1. Elemental silver, for example, nanocrystalline particles or foil.
2. Inorganic compounds/complexes, e.g., silver nitrate, silver sulfadiazine, silver oxide, silver phosphate, silver chloride, or a silver-zirconium compound.
3. Organic complexes, for example, colloidal silver preparations and silver proteins.

Colloidal silver solutions were the most common delivery system prior to 1960. They take the form of charged pure silver particles (3–5 ppm) held in suspension by a small electric current, where the positively charged ions repel each other, hence making it possible for the particles to remain in solution when applied topically to a wound. Although it is highly bacteriocidal with no resistance, colloidal silver offers little practical value since the solutions are unstable when exposed to light. When the silver ions are complexed to proteins in solution, they become more stable. However, they are also much less antibacterial than pure ionic silver.

In the 1960s various silver salts were developed. When silver ions are complexed in AgCl, $AgNO_3$, Ag_2SO_4, etc., they become more stable delivery systems. Silver nitrate was the most widely used compound, but it is dangerous to use in concentrations exceeding 2%. A 0.5% aqueous silver nitrate solution became the standard solution for treating burns and infected wounds. However, nitrate is toxic to wounds and cells, and appears to decrease healing. Silver nitrate is also unstable when exposed to light.

A number of studies have shown that silver ions and radicals produce the best antimicrobial results as well as optimizing the wound-healing environment. As a consequence, the silver salts and complexes used today were developed to maintain a sustained release of silver ions. A typical silver-containing compound is AlphaSan RC5000 developed by Milliken. As a zirconium-phosphate-based ceramic ion-exchange resin containing silver, AlphaSan is effective against a range of microorganisms that can cause undesirable effects. It is widely used in the Europe, Japan, and the United States, and has been approved by the US Food and Drug Administration for contact applications.

Table 11.1 Silver compounds used in silver-containing wound dressings

Manufacturer	Product name	Silver compound used
Argentum	Silverlon	Metallic silver
Smith & Nephew	Acticoat	Nanocrystalline silver
Medline	SilvaSorb	Silver chloride
Coloplast	Contreet	Silver-ammonium complex
Johnson & Johnson	Actisorb Silver 220	Silver carbon
Coloplast	SeaSorb Ag	Silver sodium hydrogen zirconium phosphate
Smith & Nephew	Flamazine	Silver sulfadiazine
Hollister woundcare	Restore contact layer	Silver sulfate

Source: Advanced Textiles for Woundcare (Chapter 7: Antimicrobial textiles in managing wound infection). Woodhead Publishing, 2009.

Table 11.1 summarizes the various silver compounds used in some of the many silver-containing wound dressings.

11.4.3 Methods of adding silver into wound dressings

The silver-containing wound dressings currently available have considerable differences in their overall structures, the concentration and formulation of the silver compounds, and the base materials used to carry the silver compounds. Many methods have been developed to incorporate silver into wound dressings. In general, these methods can be classified into the following four types.

1. Physical treatment of the base material: In this method, the fiber or fabric can be coated with metallic silver.
2. Chemical treatment of the base material: In this method, the fiber or fabric can be treated with silver-containing solutions, whereby silver ions can be attached to the wound dressing through ion exchange.
3. Blending: Fine particles of silver compounds can be blended with the base material.
4. Blending of silver-containing fibers with other fibers: This method is used in the production of Silvercel, where calcium alginate fibers are blended with silver-coated X-static fibers.

Because of the variations in their intended purposes, different types of silver-containing wound dressings have considerably different silver contents. Table 11.2 shows the silver contents in some silver-containing wound dressings.

11.4.4 Applications of modern silver-containing antimicrobial wound dressings

Modern silver-containing wound dressings are now widely used in the management of infected wounds. Depending on the level of silver content, these products can be divided into three main types.

Table 11.2 Silver contents in various silver-containing wound dressings

Product name	Ag content (mg/100 cm^2)
Silverlon	546.0
Acticoat	105.0
Contreet foam	85.0
Contreet hydrocolloid	32.0
Aquacel Ag	8.3
SilvaSorb	5.3
Actisorb Silver 220	2.7

Source: Advanced Textiles for Woundcare (Chapter 7: Antimicrobial textiles in managing wound infection). Woodhead Publishing, 2009.

1. Products with a high silver content: These products have a rapid silver ion release designed for wounds with heavy exudate and bacterial colonization.
2. Products with a medium silver content: These products have a more modest silver-release pattern, where silver ion is released over several days. They are claimed to be sufficient for moderate to severe pathogenic bacterial populations. The nonsilver components of these dressings are attuned to wound-bed management, that is, exudate control, debridement of wound debris, and management of the wound environment.
3. Products with a low silver content: These products are sufficient for low-grade infections in chronic wounds, but are more appropriately used as a barrier to infection in acute wounds, burns, and surgical injuries.

Figure 11.3 shows the silver concentration in the contact solution after several types of silver-containing dressings were placed in contact with wound fluid over different periods of time. It can be seen that Silverlon and Acticoat release more silver ions than other products. Both the Silverlon and Acticoat products are coated with a high level of metallic silver, which is needed for burn wounds where the prevention of infection is an important consideration. In other types of dressings, the silver ions act as an antimicrobial agent to control bacteria growth and prevent cross-infection. For example, many types of alginate wound dressings now contain silver. When a silver-containing alginate wound dressing is placed on an exuding wound, it first absorbs exudate from the wound bed into the dressing structure before the alginate fibers swell to form a hydrogel. In the meantime, silver ions are activated and liberated into the contacting solution, exerting an antimicrobial effect on the bacteria. In this case, the level of silver release can be significantly lower than in burn management.

Lansdown et al. made a sequential microbiological examination of wound swabs, sampling of wound exudate, and wound scale from seven patients with chronic wounds. After analyzing the silver content using atomic absorption spectrometry, they found that:

- all silver released into the wound bed was absorbed by wound exudate or debris (wound scale, etc.);
- silver uptake by wound exudate is approximately proportional to its viscosity (protein content);

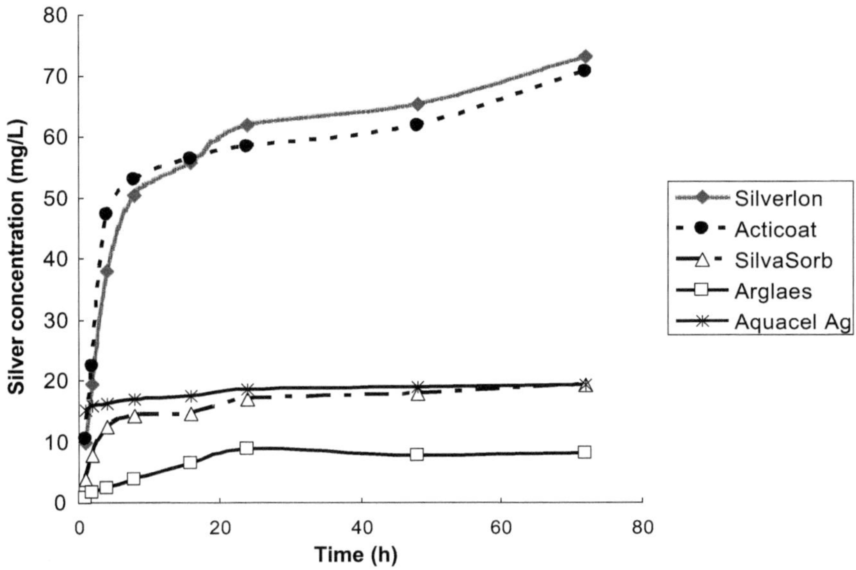

Figure 11.3 Silver concentration in the contact solution over different periods of time. *Source:* Advanced Textiles for Woundcare (Chapter 7: Antimicrobial textiles in managing wound infection). Woodhead Publishing, 2009.

- the amount of silver released from dressings is closely related to the amount of moisture absorbed;
- wounds treated with silver do not attain a germ-free status, suggesting that silver-resistant organisms such as *S. aureus* and *Pseudomonas aeruginosa* may contribute to indolence in healing.

In a detailed study on the performance of various silver-containing wound dressings, Thomas found that when testing against *S. aureus*:

- Acticoat exhibited a marked bactericidal effect within two hours;
- Contreet-H had an inhibitory effect;
- Actisorb Silver 220 appeared to prevent the proliferation of the organism within the dressing after a minimum contact time of 4 h.

When used on infected wounds, Actisorb Silver 220 removes the microorganisms from the suspension and binds them to the surface of the charcoal fibers. These microorganisms remain viable for many hours, until they are progressively inactivated by the silver ions in the dressing.

While the total silver content is important, other factors also influence a dressing's ability to kill microorganisms. These include the distribution of the silver within the dressing, for example, whether it is present as a surface coating or is dispersed through the structure; its chemical and physical form, for example, whether it is present in a metallic, bound, or ionic state; and the dressing's affinity for moisture, which is a prerequisite for the release of active agents in an aqueous environment. Products in which the silver content is concentrated on the dressing surface rather than "locked up" within its structure perform well, as do those in which silver is present in the ionic form.

Yin et al. compared the antimicrobial activity of Acticoat with silver nitrate, silver sulfadiazine, and mafenide acetate to determine their minimum inhibitory concentration

(MIC), minimum bactericidal concentration (MBC), and zone of inhibition. Although mafenide acetate produced the greatest zone of inhibition, the MBC of the product was higher than its MIC, indicating that it had a bacteriostatic rather than a bactericidal action. In contrast, the MICs and MBCs of the silver-containing products were very similar, indicating that their activity is essentially bactericidal.

11.4.5 Cost effectiveness of silver-containing wound dressings

Cost effectiveness is a practical issue with silver-containing wound dressings, since they are generally more expensive than conventional ones. However, a thorough assessment of the cost effectiveness of a healthcare intervention is complicated and involves many factors, including resource use, quality-of-life issues, and economic parameters such as the ability to work. Silver-containing wound dressings have been found to be cost effective for a number of reasons, such as:

- reduced time to wound healing;
- shorter hospital stays;
- reduced dressing-change frequency;
- reduced need for pain medication during dressing change;
- fewer MRSA bacteremias resulting from MRSA-infected wounds.

A number of studies have found that the overall costs of treating burn wounds were lower with silver-containing wound dressings in comparison to traditional treatment with silver sulfadiazine. In practice, however, healthcare reimbursement is compartmentalized and the costs of clinician time are kept separate from resource costs. This means that even if a dressing is shown to save money overall by reducing time to healing, hospital stay, or nursing time, controllers of dressing budgets may choose to restrict reimbursement to simple low-cost dressings.

11.4.6 Side effects of silver-containing wound dressings

Argyria is a condition caused by improper exposure to silver compounds. The application of silver-containing wound dressings can result in an accumulation of silver or silver sulfide particles in the skin, which darken on exposure to sunlight. This skin discoloration or staining is harmless and usually reversible. It is not true systemic argyria, which is rare and usually related to oral ingestion of silver solutions. True argyria is the result of deposition of silver compounds on the skin and internal organs, and presents as generalized blue-grey skin discoloration, particularly in light-exposed areas. True argyria is unsightly and irreversible, but usually not pathological or life threatening. The total amount of silver required to cause argyria is unknown, but total body contents of 3.8–6.4 g have been suggested.

In practice, only a small proportion of the silver presented to a wound site in a dressing is involved in antimicrobial action. Most remains within the dressing or binds to proteins in the wound or wound debris. Very little is systemically absorbed. Even if absorbed systemically, silver is excreted out of the body, mainly via the biliary route in feces. Some is also excreted in urine. Silver is not absorbed into the central or peripheral nervous systems.

Clinically, silver dressings are mainly used in the treatment of infected wounds or wounds at a high risk of infection or reinfection. In the following situations, silver dressings should not be used.

- In the absence of signs of localized, spreading, or systemic infection.
- Clean surgical wounds at a low risk of infection, for example, donor sites and closed surgical wounds.
- Chronic wounds undergoing a normal healing process.
- Small acute wounds at low risk of infection.
- Patients who are sensitive to silver or any of the dressing components.
- Wounds being treated with enzymatic debridement.
- During pregnancy or lactation.
- When contraindicated by the manufacturer—for example, some manufacturers recommend that their silver dressings are not used during magnetic resonance imaging, or on/near body sites undergoing radiotherapy.

As more and more silver-containing wound dressings are used, concerns have been raised that silver released into the environment may be harmful. In this respect, silver is used worldwide in a wide range of technologies and the environmental impact of silver is not clear. However, compared to silver released from other sources, wound dressings represent only a very small proportion of global annual silver consumption, accounting for about 0.0008%.

11.5 Other types of antimicrobial materials

In developing antimicrobial dressings to manage wound infection, it should be noted that since wound dressings are directly in contact with a broken skin, there are strict requirements for the type of applicable antimicrobial materials. In particular, these materials should meet the following requirements.

1. Are nontoxic and safe to use on a broken skin.
2. Should not cause allergic reaction.
3. Would not develop bacterial resistance.
4. Possess a broad-spectrum antimicrobial effect.
5. Have a sustained antimicrobial effect during the material's useful time.

Organic antimicrobial compounds commonly used for antimicrobial textiles have limited use in wound management, since in most cases they are not suitable for use on broken skin. In recent years two classes of materials have been increasingly used in the wound-management industry as novel materials for the manufacture of wound dressings with antimicrobial function. These are briefly discussed below.

11.5.1 Metal ions with antimicrobial properties

In recent years, silver has been successfully used as an effective antimicrobial agent and silver containing wound dressings are now widely used throughout the world. In addition to silver, other metal ions have also proven effective in preventing bacterial growth. In particular, zinc and copper ions, while being nontoxic to humans, can be easily attached to wound dressings through chelation with fibers containing amine groups. In this respect, chitosan fibers treated with zinc and copper compounds have the combined antimicrobial properties of the chitosan as well as the metal ions. Experimental results have shown that zinc- and copper-containing chitosan fibers possess excellent antimicrobial efficacy.

11.5.2 Naturally occurring antimicrobial materials

Melaleuca alternifolia oil, also called tea tree oil, has demonstrated promising efficacy in treating wound infections. Tea tree oil is antimicrobial, antiinflammatory, and has demonstrated its ability to activate monocytes. Clinical studies have shown that tea tree oil can be used effectively as adjunctive therapy in treating osteomyelitis and infected chronic wounds.

Other types of naturally occurring compounds have also demonstrated their antimicrobial efficacy. For example, honey can be used for managing wound infections; mastic gum can be used for *helicobacter pylori* gastric ulcers; and cranberry juice can be used for urinary-tract infections. Safe and effective antimicrobial wound dressings can be developed by the appropriate incorporation of these novel types of naturally occurring compounds.

11.6 Summary

The porous nature of textile-based wound dressings is ideal for bacterial colonization, and there is also a risk for the transmission of bacteria through shedding fibrous particles. Fibers with antimicrobial properties are therefore particularly useful in the production of wound dressings which can resist or control infection. While many types of antimicrobial agents have been used in the wound-management industry, silver-based antimicrobial materials have been widely used in recent years for the production of dressings that can be used to manage wound infection and cross-infection in hospital wards, although many side effects of the various forms of silver compounds still need to be thoroughly investigated.

Sources for further information

[1] Sussman C, Bates-Jensen B. Wound care: a collaborative practice manual for health professionals. Oxford: Blackwell Publishing; 2012.
[2] Leaper DJ, Harding KG, editors. Wounds: biology and management. Oxford: Oxford University Press; 1998.
[3] Kingsley A. A proactive approach to wound infection. Nurs Stand 2001;15(30):50–4.
[4] National Audit Office. The management and control of hospital acquired infection in acute NHS trusts in England. London: Stationery Office; 2000.

[5] Duckworth G, Cookson B, Humphreys H, et al. Revised guidelines for the control of methicillin-resistant *Staphylococcus aureus* infection in hospitals. J Hosp Infect 1998;39(4):253–90.
[6] National Audit Office. Improving patient care by reducing the risk of hospital acquired infection: a progress report. London: Stationery Office; 2004.
[7] McDonnell G, Russell AD. Antiseptics and disinfectants: activity, action, and resistance. Clin Microbiol Rev 1999;12(1):147–79.
[8] Molan PC. The role of honey in the management of wounds. J Wound Care 1999;8(8):415–8.
[9] Klasen HJ. Historical review of the use of silver in the treatment of burns. I. Early uses. Burns 2000;26(2):117–30.
[10] Fox C. Topical therapy and the development of silver sulphadiazine. Surg Gynecol Obstet 1968;157:82–8.
[11] Lansdown ABG. Physiological and toxicological changes in the skin resulting from the action and interaction of metal ions. CRC Crit Rev Toxicol 1995;25:397–462.
[12] Ovington LG. Nanocrystalline silver: where the old and familiar meets a new frontier. Wounds 2001;13(Suppl. B):5–10.
[13] Modak SM, Fox CL. Binding of silver sulfadiazine to the cellular components of *Pseudomonas aeruginosa*. Biochem Pharmacol 1973;22:2391–404.
[14] Wells TN, Scully P, Paravicini G, et al. Mechanisms of irreversible inactivation of phosphomannose isomerases by silver ions and flamazine. Biochemistry 1995;34: 7896–903.
[15] Demling RH, DiSanti L. Effects of silver on wound management. Wounds 2001;13(Suppl. A):5–15.
[16] Thomas S, McCubbin P. An in vitro analysis of the antimicrobial properties of 10 silver-containing dressings. J Wound Care 2003;12(8):305–8.
[17] Lansdown AB, Williams A. How safe is silver in wound care? J Wound Care 2004;13(4):131–6.
[18] Lansdown ABG, Williams A, Chandler S, et al. Silver absorption and antibacterial efficacy of silver dressings. J Wound Care 2005;14(4):155–60.
[19] Wright JB, Lam K, Hansen D, et al. Efficacy of topical silver against fungal burn wound pathogens. Am J Infect Control 1999;27:344–50.
[20] Yin HQ, Langford R, Burrell RE. Comparative evaluation of the antimicrobial activity of ACTICOAT antimicrobial barrier dressing. J Burn Care Rehabil 1999;20:195–200.
[21] Halcon L, Milkus K. *Staphylococcus aureus* and wounds: a review of tea tree oil as a promising antimicrobial. Am J Infect Control 2004;32(7):402–8.
[22] Caruso DM, Foster KN, Blome-Eberwein SA, et al. Randomized clinical study of Hydrofiber dressing with silver or silver sulfadiazine in the management of partial thickness burns. J Burn Care Res 2006;27(3):298–309.
[23] Paddock HN, Fabia R, Giles S, et al. A silver impregnated antimicrobial dressing reduces hospital costs for pediatric burn patients. J Paediatr Surg 2007;42(1):211–3.
[24] Cavanagh MH, Burrell RE, Nadworny PL. Evaluating antimicrobial efficacy of new commercially available silver dressings. Int Wound J 2010;7(5):394–405.
[25] Dimikakos E, Katsenis K, Kalemikerakis J, et al. Infected venous leg ulcers: management with silver-releasing foam dressing. Wounds 2009;21(1):4–8.
[26] Russell L. The CONTOP multinational study: preliminary data from the UK arm. Wounds UK 2005;1:44–54.
[27] Cutting K, White R, Edmonds M. The safety and efficacy of dressings with silver: addressing clinical concerns. Int Wound J 2007;4(2):177–84.

[28] Kwon HB, Lee JH, Lee SH, et al. A case of argyria following colloidal silver ingestion. Ann Dermatol 2009;21(3):308–10.
[29] Wilkinson LJ, White RJ, Chapman JK. Silver and nanoparticles of silver in wound dressings: a review of efficacy and safety. J Wound Care 2011;20(11):543–9.
[30] Leaper DJ. Silver dressings: their role in wound management. Int Wound J 2006;3:282–94.
[31] Qin Y. Silver containing alginate fibers and dressings. Int Wound J 2005;2(2):172–6.
[32] Hooper SJ, Percival SL, Hill KE, et al. The visualisation and speed of kill of wound isolates on a silver alginate dressing. Int Wound J 2012;9(6):633–42.
[33] Moore Z. Top tips on when to use silver dressings. Wounds Int 2013;4(1):15–8.

Medical textile products for the control of odor

12

12.1 Introduction

Odor or odor (commonly referred to as smell) can be described as the property of a substance that is perceptible by the sense of smell. An odor is caused by one or more volatilized chemical compounds, generally at a very low concentration, that humans or animals perceive by the sense of olfaction. Odors are also commonly called scents and can be pleasant or unpleasant. The terms fragrance and aroma are used primarily by the food and cosmetic industries to indicate a pleasant odor, and are sometimes used to describe perfumes. In contrast, malodor, stench, reek, and stink are used specifically to describe unpleasant odors.

Odors from the human body come from secretions from the sweat glands, urine, feces, expiration, saliva, breasts, skin, and sex organs. Healthy individuals can emanate strong odors due to a range of factors, such as their personal physiology, diet, physical activities, and hygiene practices. An unhealthy individual may suffer an increase in unpleasant and strong body odor due to an imbalance in personal microflora, a difference in metabolic activity, problems with urinary or fecal incontinence, and from the ingestion of therapeutic medicines. Malodor may also arise from wounds which have become infected or have necrotic and dying tissue. People emanating strong odors as a result of infected or dying tissue, incontinence, or disease can become withdrawn, isolated, and depressed as the malodors impact on their relationships with family and friends.

Body odor is influenced by the actions of the skin flora, including members of *Corynebacterium*, which generate enzymes called lipases. The lipases are able to break down the lipids in sweat to produce smaller molecules such as butyric acid. Propionic acid is a breakdown product of some amino acids by propionibacteria, which thrive in the adolescent and adult sebaceous glands. Because propionic acid is chemically similar to acetic acid, with similar characteristics, body odors may be identified as having a vinegar-like smell by certain people. Isovaleric acid is the other source of body odor, as a result of actions by *Staphylococcus* epidermidis.

Since the exuding wounds contain a cocktail of proteins, carbohydrates, enzymes, and other ingredients that are closely associated with odor production, they are often a source of obnoxious smells, especially when infection is involved. Wound odors are caused by short-chain organic acids such as *n*-butyric, *n*-valeric, *n*-caproic, and *n*-caprylic acids produced by anaerobic bacteria, and amines and diamines such as cadaverine and putrescine that are produced by the metabolic processes of other proteolytic bacteria. Organisms frequently isolated from malodorous wounds include anaerobes such as *Bacteroides* and *Clostridium* species, and numerous aerobic bacteria including *Proteus*, *Klebsiella*, and *Pseudomonas*.

Medical Textile Materials. http://dx.doi.org/10.1016/B978-0-08-100618-4.00012-1

Strong odors arising from the skin and wounds, or as a result of incontinence, are problematic and can be unpleasant for both patients emitting odors and those in close proximity with them, such as carers, family members, etc. To solve this problem, textile materials can be used to control odor through a number of mechanisms, such as absorption of odorous volatiles, release of antimicrobial agents, and dispersal of scents.

12.2 Qualitative description of wound and body odor

Because of the many chemicals involved in odor generation, smells vary greatly from pleasant to obnoxious. There are many adjectival words to describe smells, including acrid, antiseptic, bitter, burning, choking, clean, delicious, fragrant, fresh, medicinal, musty, pungent, putrid, rancid, rich, rotten, salty, smoky, sour, spicy, stale, stinky, strong, sweet, etc. Even so, being highly subjective, descriptive analysis of odor is problematic since the panels of assessors may disagree on the terms used to quantify and qualify odors, which can result in confusing and ambiguous findings. Because the sense of smell by an individual is highly subjective, terms such as "typical" or "classic" odor have been applied in many cases.

Since the characterization of odor is critical to the technical development of medical textile materials for odor control, it is important to have a quantitative assessment method that can offer a detailed classification of body odor. In this respect, many attempts have been made to establish qualitative description and quantitative grading of body odor. For example, six descriptive terms can be used to describe axillary sweat extracts: acid, chicken broth, onion, sweat (acrid), butter, and floral. The intensity of the different types of smells can be graded on an 11-point intensity scale from 0 (none) to 10 (very strong), using which a highly trained panel of assessors can carry out a reasonably accurate assessment of body odors.

12.3 Control of unpleasant body odor

Body odor presents a social stigma and causes embarrassment if not taken care of properly. As a way of life, body odor can be controlled by having a healthy lifestyle, for example, by bathing regularly, wearing clean clothes, applying deodorant, eating proper food, shaving excess body hair, etc. From a scientific point of view, the first consideration should always be given to the source of the problem. If the odor is generated by bacteria, oral or topical medication should be applied to reduce microbial infection.

It is well recognized that certain types of wounds produce obnoxious odors which, even in moderate cases, can cause significant distress or embarrassment to patients and their relatives. In extreme cases this odor can become so overpowering that it may cause an individual to withdraw from social contacts, even with family and close friends. With malodorous wounds, the most effective method is to prevent or eradicate the infection responsible for the odor. This may be achieved in a number of ways. The administration of systemic antibiotics or antimicrobial agents may be effective in some cases, but it is often difficult to achieve an effective concentration of the antibiotic at the site of infection, particularly in the presence of slough or necrotic tissue. Topical antiseptics are not usually applied on wounds, since many have adverse effects on wound healing. One preparation that has been found to be effective in certain situations is a hydrogel containing a suitable concentration of metronidazole, typically about 0.8% w/v. This concentration is effective against a range of aerobic organisms responsible for odor generation.

If it is not possible to eliminate the bacteria responsible for production of the odor, it then becomes necessary to deal with the problem by some other means. Historically wound odors were masked by burning incense, and in more recent times by the use of aerosols or air fresheners. Although these do not resolve the underlying problem, they may make life a little more bearable for the patient and the family. In recent years activated carbon materials have been widely used to absorb odor. A typical example is Actisorb 220 from Johnson & Johnson, which is a composite product containing activated carbon to absorb odor and silver ions to kill bacteria. Other textile-based materials can also be utilized to reduce the intensity of malodor through the absorption of volatile odors or masking the malodor by dispersing fragrances.

12.4 Medical textile materials with odor-control properties

Textile materials are usually in close contact with the human body, and the relationship between body odor and textiles can be understood in two ways: first, the penetration and build-up of odor within the textile structure due to sweat or other human secretions or excrement in bedding and clothing; and second, the control of odor through the use of textiles, for example, bandages and dressings, incontinence products, flatus devices, etc. During these processes, the interaction of textile materials with various components from the body can significantly influence the odor-absorbing performance. For example, many absorbent textiles are composed of inherently hydrophilic fibers, which can absorb odorous compounds. On the other hand, the adsorption of lipids to the surfaces of hydrophobic fibers such as polyester or polypropylene can provide an excellent source of nutrients for some bacterial strains, which in turn become a source of odor generation.

In essence, an odor-management strategy considers both the cause of odor generation and the management of odor with textile-based materials. Firstly, odor control can be achieved through the use of antimicrobial materials to control the growth of bacteria and fungi. Secondly, absorbent materials can be incorporated into textiles to absorb malodor and odorous volatiles. Thirdly, textile materials can be used to disperse scents to mask the malodor. These methods are described below.

12.4.1 Antimicrobial textiles

Antimicrobial agents can be applied to fibers, yarns, and textile fabrics to kill or inhibit the growth of microorganisms such as bacteria, fungi, yeasts, and viruses, which colonize on or inside the textile structure. Most antimicrobial agents used in textile materials are biocidal and act by damaging the cell wall, altering the cell membrane permeability, denaturing proteins, or inhibiting and altering essential functions of the microorganisms' metabolic pathways. Common antimicrobial agents include silver compounds, tricoslan, polyhexamethylene biguanide, and quaternary ammonium compounds.

Traditionally, antimicrobial agents are incorporated into a textile material to inhibit the growth of or kill microorganisms within the textile structure, rather than to influence the resident or nonpathogenic skin microflora of the person wearing/using the textile. Therefore, unless the antimicrobial agent leaches from the textile material, complete control of human body odors may not be possible since the treated textile will only control odor development within the textile and not that developed on the skin or other areas of the body. Nonetheless, reducing odor build-up within textiles may still be necessary, since odor originating on the body and odor emitted from textiles in contact with the body can differ in quality and intensity. These differences can result from variations in humidity, temperature, airflow, nutrients, and the presence of antimicrobial agents on the skin and textile.

The incorporation of antimicrobial agents into wound dressings has long been used to reduce bacterial load in the wound and/or reduce the adherence and growth of bacteria on the wound dressing itself. The reduction of microbial loads in wounds is beneficial to the process of wound healing, and at the same time helps to lower the level of microorganisms responsible for the generation of wound malodor. Silver is the most common antimicrobial agent incorporated into wound dressings, while iodine-based gauzes or dressings are also effective in controlling bacterial growth.

In recent years tea tree oil has found widespread uses as an antimicrobial agent. It is an essential oil with a fresh camphoraceous smell and a color that ranges from pale yellow to nearly colorless and clear. Tea tree oil is extracted from the leaves of *Melaleuca alternifolia*, which is native to southeast Queensland and the northeast coast of New South Wales, Australia. When incorporated into textile materials, its antimicrobial effect is effective in a variety of dermatologic conditions, including dandruff, acne, lice, herpes, and other skin infections.

12.4.2 Odor-absorbing textiles

The structure of textile materials can be engineered to have odor-absorbing properties. For example, cyclodextrins can be added into a textile material to enhance its odor-absorbing capabilities. As shown in Figure 12.1, cyclodextrins are cyclic oligosaccharides which are made up of six to eight glucopyranoside units. They are capable of forming inclusion complexes with other molecules due to their toroidal shape, which is largely hydrophobic. Malodors can be controlled through the use of cyclodextrins by absorbing odorous molecules into their internal structure. In particular, incorporation

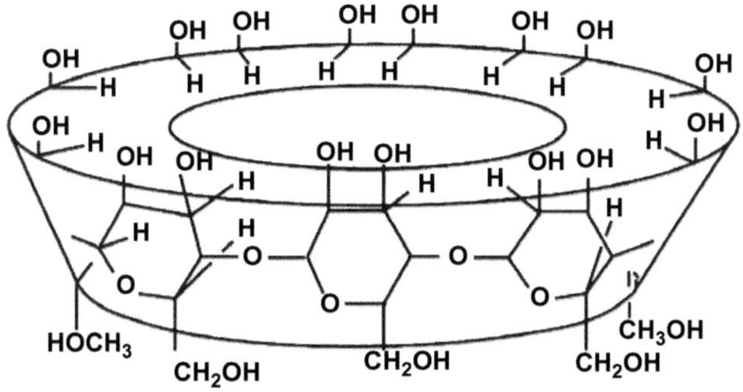

Figure 12.1 The chemical structure of cyclodextrin.

of cyclodextrins into wound dressings is effective in absorbing butyric acid and valeric acid odorants. As well as cyclodextrins, other odor-absorbent materials that can be used for feminine hygiene products, disposable diapers, and incontinence products are zeolites, silicates, absorbing gelling materials, starches, ion-exchange resins, etc.

12.4.3 Scented textiles

Humans are highly sensitive to the influence of odors in their living environment, and human emotions such as relaxation, exhilaration, sensuality, happiness, and well-being are closely related to odors that stimulate the olfactory pathways in the brain and, in particular, the limbic system. Throughout the world, the application of fragrances to mask the presence of unpleasant odors has been a common method employed to refresh a domestic environment. Personal hygiene products and cosmetics are frequently scented in order to mask natural odors emanating from the body. Perfumed substances in the form of an aerosol, liquid, or solid are often sprayed or rubbed directly on to the skin.

In a similar way, scents can be applied on textiles and nonwoven fibrous hygiene products. However, since perfumes are volatile, the intensity of the fragrance will dissipate quickly. It is a challenge to make the fragrances incorporated into textiles provide long-lasting scent. It has been found that by encapsulating the liquid or gas fragrance in a small capsule, the storage life of the volatile compound can be extended to offer a slow release of the scents over time.

In the development of cosmetotextiles, a process of microencapsulation is frequently used to incorporate bioactive ingredients that are designed to release over time. In particular, cosmetotextiles are often impregnated with a finish composed of solid microcapsules, each holding a specific amount of cosmetic substance meant to be released to the skin surface. Similarly, a new dermotextile technology uses natural-based microparticles as cosmetic carriers instead of microcapsules, which can be safer and healthier for medical applications. These novel techniques can be used to carry jasmine, lavender, and other aromatherapy compounds to mask unpleasant odors in a nursing home. In addition, by using cyclodextrins to incorporate rosemary, lavender, jasmine, lemon, and sandalwood on cotton, it is possible for the textile material to emanate scents over a long period of time. Alternatively, the scent molecules can be infused into the polymer chips prior to spinning, so that fibers containing the scent molecules can be manufactured to offer long-lasting fragrances.

12.5 Activated carbon for odor control

In the wound-management industry where odor control is much required, a new approach was adopted in 1976, when Butcher et al. first described the use of an activated charcoal cloth developed by the Chemical Defence Establishment in Porton Down, UK. This material was incorporated into pads containing surgical gauze and water-repellent fabric. When these pads were used in the treatment of fungating breast cancer, gangrene, and colostomies, the associated odors were said to be totally suppressed.

Activated charcoal cloth is produced by carbonizing a suitable cellulosic fabric by heating it under carefully controlled conditions. The production process can be carried out in two ways. In the physical reactivation process, the precursor is developed into activated carbons by using gases in one or a combination of the following processes.

1. Carbonization: Material with carbon content is pyrolyzed at temperatures in the range of 600–900 °C, in the absence of oxygen (usually in an inert atmosphere with gases like argon or nitrogen).
2. Activation/oxidation: Raw material or carbonized material is exposed to oxidizing atmospheres (carbon dioxide, oxygen, or steam) at temperatures above 250 °C, usually in the range of 600–1200 °C.

In the chemical activation process, prior to carbonization the raw material is impregnated with certain chemicals. The chemical is typically an acid, a strong base, or a salt, such as phosphoric acid, potassium hydroxide, sodium hydroxide, calcium chloride, or zinc chloride. Then the raw material is carbonized at temperatures in the range of 450–900 °C. It is believed that the carbonization/activation step proceeds simultaneously in the chemical activation method. Generally speaking, chemical activation is preferred over physical activation due to the lower temperatures and shorter time needed.

During the activation process the surface of the carbon breaks down to form small pores, which greatly increase the effective surface area of the fibers and hence their ability to remove unpleasant smells, as it is believed that the molecules responsible for the production of the odor are attracted toward the surface of the carbon and held there by electrical forces. The large surface area of the activated carbon enables it to take up very large numbers of molecules, making it possible to remove odor over a prolonged period. It is estimated that a gram of activated carbon can have a surface area in excess of 500 m^2, reaching 1500 m^2 in some cases.

Since 1976 a number of odor-controlling dressings containing activated charcoal have been produced commercially. The first of these was Actisorb from Johnson & Johnson. In laboratory studies it was found that if the dressing was shaken with a suspension of bacteria, the organisms became firmly attached to the charcoal fabric and were thus removed from the solution, although they still remained viable. A modified version of this dressing, Actisorb Plus, was then designed to contain 0.15% silver, which was effective in killing the bacteria bound on to the charcoal cloth.

A number of other activated carbon dressings have also been developed. These products vary in structure and composition, and hence in their ability to cope with wound exudate, which is an important feature of malodorous wounds. Some of these products are placed directly on the wound surface, while others are designed as secondary dressings placed over a primary dressing but beneath the retaining dressing or bandage. For example, CarboFlex from ConvaTec is a multicomponent, nonadhesive dressing with an absorbent wound contact layer. The activated charcoal cloth is positioned behind an absorbent layer made of a mixture of alginate fibers and carboxymethyl cellulose fibers, and a perforated plastic film which provides an unidirectional flow of liquid. Another absorbent layer and plastic film are positioned behind the activated charcoal fibers.

There are many types of wound dressings containing activated carbon as the odor-absorbing component. Carbonet from Smith & Nephew is a primary dressing which can be used with or without a secondary dressing. It is a multicomponent dressing with a wound contact layer composed of knitted viscose rayon which is backed by an absorbent cellulosic layer, and bonded to activated charcoal cloth sandwiched between the two layers of polyethylene net. CliniSorb from CliniMed is an activated charcoal cloth sandwiched between the layers of viscose rayon. It is recommended to be used as a secondary dressing. Lyofoam C from Mönlycke is a direct-contact dressing held in place with additional tape or bandage. A layer of activated charcoal cloth is held between a hydrophilic polyurethane contact layer and a hydrophobic polyurethane outer layer. This product has good fluid-handling capabilities since the polyurethane sheets can absorb a large amount of wound exudate. Pharmapad from Pharma-plast is a secondary dressing composed of three layers, with activated charcoal cloth sandwiched between the two layers of viscose/nylon "thermal bond" nonwoven. Sorbsan Plus Carbon from Unomedical is a multilayer activated charcoal dressing with a calcium alginate skin-contact layer combined with a secondary absorbent layer.

12.6 Characterization of odor-absorbing products

The main property of odor-absorbing products is their ability to remove volatile chemicals responsible for a smell. From a technical point of view, the characterization of the absorption property is concerned with the design of testing apparatus and the quantitative or qualitative testing of either the smell or the chemicals responsible for smell generation. In addition, the many diversified smells require accurate though still subjective description. Some of the key aspects concerning test methods used for the characterization of odor-absorbing products are introduced below.

12.6.1 Subjective description of odor

Odor is detected through the sense of smell, and the human nose is the most effective instrument for assessing objectionable odors. In a hospital or other care facility, patients, health practitioners, friends, and family members may all be able to detect the presence of unpleasant odors emanating from a patient's body, wounds, or soiled bedding or clothing. Since the odor-absorbing property is mainly a subjective reflection of a product's performance, the perception of a healthy assessor is one of the most important test results regarding the qualities of odor-absorbing products.

There are many subjective test methods. For example, discrimination testing is used to determine whether there is a detectable difference among two or more products. The test uses a trained panel to discriminate from one product to another. Examples of discrimination tests are paired comparison tests, triangle tests, duo-trio tests, etc. Category scales involve a method of rating a stimulus by assigning a value to it on a scale of 0–10 to represent a worsening smell from pleasant to obnoxious. The scale numbering is important, in that an insufficient number of categories could render the scale unsuitable for distinguishing among many levels of stimuli, whereas too many points may increase the likelihood of detecting differences which are not actually there.

Line scales or visual analog scales require assessors to rate a sensation by placing a mark on a line which has been anchored at each end point by some form of descriptor (e.g., weak/strong, low/high). A perceived advantage of the line scale over the category scale is that assessors are less restricted in their choices, as the line scale offers a continuously graded choice of alternatives and potentially greater discrimination can be obtained.

Scaling is useful in sensory measurement when there is a need to quantify sensations. Both category and line scales are useful in that they are suitable for diverse applications, simple to use, and easy to analyze statistically. Both are cost and time effective, as they allow at least one data point for a single stimulus.

In addition to these descriptive methods, ranking is a process of arranging three or more samples in order based on some selected attribute (e.g., high to low), and is generally a forced-choice procedure where ties are not allowed. The resulting ranks offer no measure of the size of difference between/among samples, and as straight comparisons are made among samples, there is no effective means of comparing samples from one session to another.

Magnitude estimation is the process of assigning values to the intensities of an attribute in such a way that the ratios between assigned values are the same as those between the magnitudes of the perceptions to which they correspond. As judgments are expressed in proportions, one scale can be compared directly with another. By using a common reference sample for each assessor, judgments are likely to be more reliable than those that may be influenced by some type of cross-comparison operative in category or line scaling.

12.6.2 Objective tests for odor-generating chemicals

Sensory evaluation is a scientific discipline used to evoke, measure, analyze, and interpret reactions to characteristics of materials/substrates as they are perceived by the senses of sight, smell, taste, touch, and hearing. The appropriate test method selection (e.g., paired comparison, scaling tests, ranking, etc.) is important, as inadequate information may be gathered from the samples under evaluation. In general, sensory measurement of odor involves the measurement of its quality and intensity, and can be carried out through instrumental evaluation using chemical or electronic sensors to determine the concentration and type of odor compounds in the air or within the textile substrate. In addition to tests for odor, antimicrobial efficacy tests are also commonly used to measure or predict the likelihood of the build-up of odor in antimicrobial-treated textiles.

The physicochemical approach to odor detection involves instrumental analysis of the chemical structure and concentration of an odorant. The relationship between the sensation of an odor as perceived by a human and the physical or chemical properties of the odor is not as simple as other sensations such as color or sound, which can be related to known physical forms such as intensity and wavelength. Measurement of odorant structure and concentration is possible through quantitative chemical analytical techniques. The ideal situation is to develop an instrument that is capable of replicating the human olfactory system without the added problems of human variability, sensitivity, and perception.

Typically, two approaches are taken to measure the presence and concentration of odorants on textiles: direct analysis of a textile substrate and analysis of the headspace above a sample. Direct analysis of the textile substrate usually involves a multistage process of extraction and clean-up before sample analysis using gas chromatography-mass spectrometry analytical techniques. Coupling gas chromatography techniques with olfactometry (GC-O) is a useful way to determine odor-active compounds in a substance. GC-O has been applied successfully for many years, albeit the still relies on a human as the odor-detecting instrument.

Analysis of the headspace above a sample is an alternative to direct measurement and chemical extraction techniques. Instrumentation for the detection of headspace volatiles includes an array of gas sensors (electronic noses), online real-time chemical ionization apparatus, etc. Some of these instrumental analysis methods are described below.

Electronic Noses. Electronic noses have been applied in the food industry for quality control and detection of contaminants and "off" odors, in environmental monitoring of livestock waste and sewage treatment works, and in the medical field for analyzing bacterial metabolites. In particular, electronic noses may be used to measure the improvement in wound odors over a period of time. Electronic noses typically consist of an array of gas-sensitive semiconductors connected to an appropriate pattern-recognition system which is able to detect complex odors. They can have up to 40 sensors, each calibrated for a different chemical specificity, which when combined provide a measurement pattern. The electronic nose relies on pattern recognition and therefore cannot identify unknown (or unexpected) compounds. Common types of chemical-sensing arrays are metal oxide sensors, conducting polymers, and piezoelectric sensors. Tin oxide detectors have many "nose" applications and are capable of detecting a large number of organic volatiles. However, high temperatures (300–400 °C) are typically required to operate the devices, which make them costly to run, and the elevated temperatures lead to a risk of the volatiles breaking down at the sensor surface. Quartz-resonator and surface-acoustic-wave devices consist of a piezoelectric substrate coated with a sensing membrane. Conducting polymers, derived from aromatic and heteroaromatic materials, measure the resistance changes in the semiconducting film.

Online Real-Time Chemical Ionization Apparatus. Instruments utilizing the concept of chemical ionization allow real-time analysis without the necessity for preconditioning or complex extraction. They avoid the fragmentation of organic molecules that occurs with electron-impact ionization. The proton transfer reaction mass spectrometer and selected ion flow tube mass spectrometer both use chemical ionization within a flow drift tube. Unlike conventional electron-impact ionization, chemical ionization does not undergo strong fragmentation, and thus it is a preferable technique for detecting volatile organic compounds when a mixture of gases is to be analyzed.

12.6.3 Testing apparatus and procedures

When assessing the odor-absorbing properties of wound-management products, Thomas et al. developed a novel apparatus for analyzing the effectiveness of five commercially available wound dressings on limiting the movement of a volatile odorous diethylamine into the surrounding environment. In their experiment, four of the five dressings contained activated charcoal (Actisorb Plus, CarboFlex, Carbonet, and Lyofoam C) and one contained no activated charcoal but was highly absorbent (Release). A test solution containing 2% diethylamine was applied to the dressings through a perforated plate at a rate of 30 mL/h in an enclosed chamber which was connected via inlet and outlet ports to a portable ambient analyzer.

During the test, the volume of test solution which was applied to the dressings before 10 ppm of diethylamine was detected in the chamber was recorded. Results showed that all dressings delayed the presence of diethylamine in the air above the dressings to some degree, including the Release dressing which did not contain activated charcoal. However, the presence of activated charcoal increased the amount of solution that could be added to the dressings compared to the absorbent dressing only. Figure 12.2 shows the performance of three wound-dressing products under the odor-absorption test.

12.6.4 Presentation of test results

As the human nose is an imperfect instrument for creating and conducting standardized tests and assessments of objectionable odors, test results are uniquely dependent on the odor acuity and consistency of the assessors. The terms "expert" or "experienced judge" have been used, often without further details on the definition of an expert. In some axillary odor trials, assessors were reportedly selected on their odor acuity. In most clinical trials where wound odor is included in the assessment of improvement of wound healing, the assessor is usually a nurse, a wound clinic practitioner, or so-called "study personnel" who presumably have expertise in caring for all aspects of wounds but may not be selected based on odor acuity. In some trials the patients themselves are required to evaluate odor, either with or without an additional evaluation by a professional.

Different types of category scales have been used to evaluate wound odor in clinical settings by either health practitioners or patients. For example, many clinical trials have used a four-point category scale comprising no odor, little, moderate, and severe

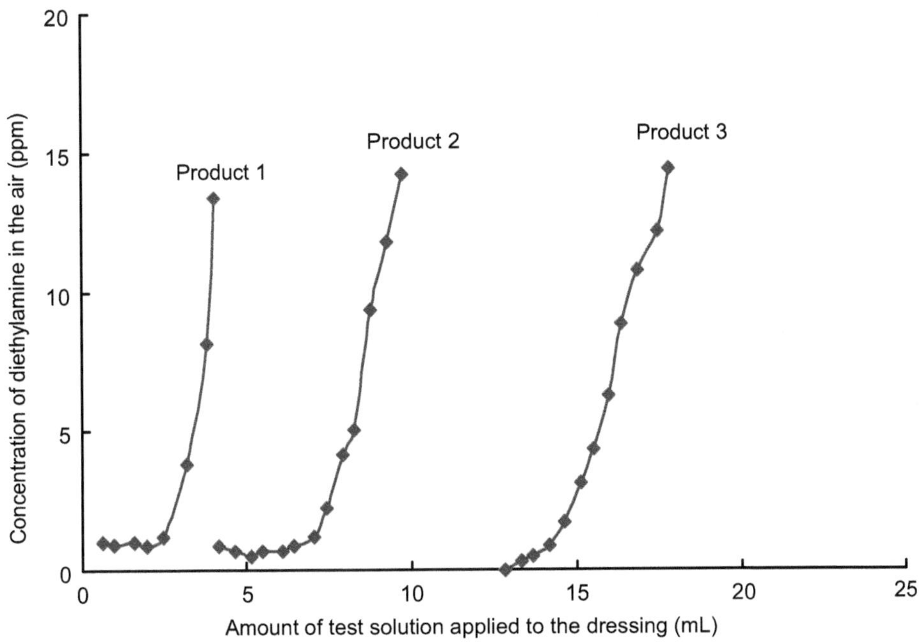

Figure 12.2 Performance of three wound-dressing products under an odor-absorption test.

odor. Another common rating scale used for the assessment of wound odor is the "odor assessment scoring tool," which was adapted from a scale originally used by Baker and Haig. The advantage of this scale for clinical use is that it provides greater detail than many other rating scales, since the assessor's presence relative to the patient and the dressing is inherent with the scale. This scale could be used for assessing improvement in odor over a treatment period, but would be less appropriate for the evaluation of odor-absorbing dressings in controlling malodor. The odor assessment scoring tool and the Baker and Haig scale use the following categories.

1. Strong: Odor is strong upon entering the room (6–10 feet from the patient) with the dressing intact.
2. Moderate: Odor is evident upon entering the room (6–10 feet from the patient) with the dressing intact.
3. Slight: Odor is evident in close proximity to the patient when the dressing is removed.
4. No odor: No odor is evident, even at the patient's bedside, with the dressing removed.

A wound-odor measurement scale offering greater discriminatory ability is a 10 cm line scale labeled numerically along the scale from 0 to 10 (0 corresponding to "no wound odor," 1–4 to "mildly offensive," 5–8 to "moderately offensive," and 9–10 to "extremely offensive"). This method can be used to monitor wound odor on a daily basis, but it can often be difficult to compare ratings across different time periods, sessions, or settings as sensory measurement can be affected by the context in which a stimulus is assessed. In this respect, the discriminative capability of a four-point scale is better suited for assessing the improvement of wound odor in clinical practice.

12.7 Summary

Odor is present in many areas of human life, and the control of unpleasant odor is an important task for healthcare practitioners as well as textile engineers. Many techniques are available for odor control, such as the use of antimicrobials in textile materials to control potentially infectious microbes that are responsible for the development of malodor, and the absorption of odorants into a textile substrate by incorporating cyclodextrins within textiles. In addition, with an increased focus on developing effective odor-controlling textiles, improved methods for objectively assessing an improvement or absence of odor are required. Owing to the complexity and diversity of odorants which can emanate from the human body and infected wounds, and because of the variability in sensitivity of the sense of smell, the development of reliable and reproducible sensory test methods for detecting the quality and/or intensity of odors is still needed. Attention should be given to the types of sensory methods used, the screening and selection of a sensory panel, and appropriate methods for collecting odor on textiles. Enhancement and refinement of the instrumental methods that can detect volatile odorous compounds in real time, at sensitivities closer to the odor thresholds, are still necessary in many applications.

Sources for further information

[1] Bartels VT, editor. Handbook of Medical Textiles. Cambridge: Woodhead Publishing Ltd; 2011.

[2] American Association of Textile Chemists and Colorists. AATCC test method 100: antibacterial finishes on textile materials. North Carolina: American Association of Textile Chemists and Colorists; 2004.

[3] American Society for Testing and Materials. ASTM E2149-10 standard test method for determining the antimicrobial activity of immobilized antimicrobial agents under dynamic contact conditions. Pennsylvania: ASTM International, PA; 2001.

[4] Ara K, Hama M, Akiba S, et al. Foot odor due to microbial metabolism and its control. Can J Microbiol 2006;52(4):357–64.

[5] Bowler PG, Davies BJ, Jones SA. Microbial involvement in chronic wound malodour. J Wound Care 1999;8(5):216–8.

[6] Greenwood JE. Monitoring wound healing by odour. J Wound Care 1997;6(5):219–21.

[7] Grey JE, Enoch S, Harding KG. Wound assessment. BMJ 2006;332:285–8.

[8] Guillet G, Zampetti A, Aballain-Colloc M. Correlation between bacterial population and axillary and plantar bromidrosis. Eur J Dermatol 2000;10:41–2.

[9] Hampson JP. The use of metronidazole in the treatment of malodorous wounds. J Wound Care 1996;5(9):421–5.

[10] Haughton W, Young T. Common problems in wound care: malodorous wounds. Br J Nurs 1995;4(16):959–63.

[11] Henkin RI. Body odor. JAMA 1995;273(15):1171–2.

[12] Holloway S, Bale S, Harding K, et al. Evaluating the effectiveness of a dressing for use in malodorous, exuding wounds. Ostomy Wound Manage 2002;48(5):22–8.

[13] Lawless HT, Heymann H. Sensory evaluation of food: principles and practices. New York: Chapman and Hall; 1998.

[14] Lee G, Anand SC, Rajendran S, et al. Efficacy of commercial dressings in managing malodorous wounds. Br J Nurs 2007;16(6):S14–20.
[15] Wang CX, Chen SL. Fragrance-release property of β-cyclodextrin inclusion compounds and their application in aromatherapy. J Ind Text 2005;34(3):157–66.
[16] Moss CW, Dees SB, Guerrant GO. Gas chromatography of bacterial fatty acids with a fused silica capillary column. J Clin Microbiol 1974;28:80–5.
[17] Thomas S, Hay NP. The antimicrobial properties of two metronidazole medicated dressings used to treat malodorous wounds. Pharm J 1991;246:264–6.
[18] Schmidt RJ, Shrestha T, Turner TD. An assay procedure to compare sorptive capacities of activated carbon dressings: the detection of impregnation with silver. J Pharm Pharmacol 1988;40:662–4.
[19] Mulligan CM, Bragg AJ, O'Toole OB. A controlled comparative trial of Actisorb activated charcoal cloth dressing in the community. Br J Clin Pract 1986;9(1):145–8.

Medical textile materials with drug-releasing properties

13

13.1 Introduction

In pharmacology, a drug is defined as a chemical substance used in the treatment, cure, prevention, or diagnosis of disease or otherwise used to enhance physical or mental well-being. Pharmaceutical drugs may be used for a limited duration or on a regular basis for chronic disorders. Recreational drugs are chemical substances that affect the central nervous system, such as opioids or hallucinogens. Alcohol, nicotine, and caffeine are the most widely consumed psychotropic drugs worldwide, and may have effects on perception, consciousness, personality, and behavior.

Textile materials are traditionally used for the comfort and general well-being of the wearer, and most textile-based products are used on the interface between the body and its external environment. Close proximity to the skin makes textile materials ideal carriers of drugs and other bioactive agents that can be slowly transferred to the skin. For this reason, transdermal delivery of drugs through the skin into the bloodstream has gained popularity in recent years. An advantage of a transdermal drug delivery route over other types of medication delivery such as oral, topical, intravenous, and intramuscular is that it provides controlled release of the medication into the patient, usually through either a porous membrane covering a reservoir of medication or body heat melting thin layers of medication embedded in the contact layer. The main disadvantage to transdermal delivery systems stems from the fact that the skin is a very effective barrier. As a result, only medications whose molecules are small enough to penetrate the skin can be delivered by this method.

Technically, the loading and release of drugs to and from textile materials are important aspects in the development and production of drug-containing textile materials. To start with, drugs can be loaded to textile fibers and fabrics via a number of methods, such as by using ion exchange, hollow fibers, and surface coating, where attention should be paid to technical issues like biostability, biodegradability, controllability, toxicity, carcinogenicity, interface reactions, material costs, and fabrication conditions. In the second step, drugs should be released in a controlled and prolonged fashion, and hence it is important to understand the various issues affecting the drug-release characteristics, such as the properties of the encapsulating material, mode of encapsulation, concentration of the drug, and the matrix architecture.

Because of the diversity in both the choice of textile materials and the requirement for specific drug deliveries, it is difficult for any drug delivery system or approach to meet all user requirements. In general, textile-based drug delivery materials have many advantages over more conventional types of drug delivery methods such as transdermal patches. These materials can be engineered to have the required rate of drug release to achieve

Medical Textile Materials. http://dx.doi.org/10.1016/B978-0-08-100618-4.00013-3

therapeutic efficacy. By selecting the appropriate base materials, textile-based drug delivery systems can be biocompatible, inert, mechanically strong, physically comfortable for the patient, capable of delivering a high payload, safe from accidental release, simple to administer, minimally off targeting, economical to fabricate, and easy to sterilize.

13.2 Development of drug delivery systems

In recent years advances in microencapsulation technologies and novel fiber-making processes have made textile fibers an attractive alternative for the encapsulation and delivery of therapeutic drugs. For example, microfiber-based medical textiles such as sutures and wound dressings can be made to contain a drug or an antimicrobial agent. Nanofibrous drug delivery systems have also been developed, and have the advantages of a large surface area to volume ratio, high porosity, and flexibility. The electrospinning technique has been modified to include multiple needles, needleless, and coaxial forms suitable for the production of drug-loaded fibrous materials which benefit from the simplicity of the electrospinning fabrication process, the diversity of electrospinnable materials, and the unique features associated with electrospun fibers. Drug-loaded electrospun nanofibers have found applications in a broad range of fields, including biomedical applications, tissue engineering, wound dressing, cosmetics, functional materials, and devices such as composite reinforcement, filters, protective clothing, and smart textiles.

Historically, in the search to improve both quality and length of life, humans have made successive developments in hygiene and nutrition that helped to increase life span. In the last few centuries the growth in medical knowledge has greatly contributed to the improvement of human health, and as a result life expectancy has increased in many parts of the world. At the same time, many attempts have been made to improve the comfort of life, and textile-based systems are particularly useful here since these products are in direct contact with the human body. The delivery of drugs by textile materials is therefore an important means to improve the quality of life for people with various medical conditions.

Use of drugs in the medical field has long been an established process, using various delivery systems for pills, ointments, injections, and such like. In oral delivery systems such as tablets, pills, and capsules the drug is absorbed in the stomach or intestinal tract, where it may lose its efficacy before being able to fulfill its medical purposes. Relatively high doses are therefore needed to achieve a medicinal effect. The delivery of drugs through the skin circumvents the passage through the liver, allowing drug dosages to be significantly reduced. In addition, transdermal delivery of drugs overcomes the many problems of traditional oral delivery systems, such as dosing children and people with swallowing difficulties.

With prolonged drug treatment, textile-based slow-release systems are preferable to daily injections or intake of pills, and textile-based drug delivery systems have the additional benefit of offering protection for the patient. Over the years, wearing comfort and other related qualities of textile products have been significantly improved, due to the continued development in textile materials and the related manufacturing technologies.

Textile materials can be made with permeable and breathing structures, and treated to have an absorptive capacity. In addition, many novel base materials have been developed to carry drugs in a fibrous structure. For example, chitosan is a partially deacetylated polymer obtained from the alkaline deacetylation of chitin, which is a glucose-based unbranched polysaccharide that occurs widely in nature as the principal component of exoskeletons of crustaceans and insects, as well as of the cell walls of some bacteria and fungi. Chitosan exhibits a variety of physicochemical and biological properties, which give it numerous applications in fields such as waste water treatment, agriculture, fabric and textiles, cosmetics, nutritional enhancement, and food processing. In addition to its lack of toxicity and allergenicity, its biocompatibility, biodegradability, and bioactivity make it a very attractive substance for diverse applications as a biomaterial in the pharmaceutical and medical fields. In another example, responsive polymer materials are playing an increasingly important part in a diverse range of applications, such as drug delivery, diagnostics, tissue engineering, and "smart" optical systems, as well as biosensors, microelectromechanical systems, coatings, and textiles. These polymers and the fibers made from them can adapt to surrounding environments, regulate transport of ions and molecules, change wettability and adhesion of different species on external stimuli, and convert chemical and biochemical signals into optical, electrical, thermal, and mechanical signals, and vice versa.

Technically, drugs can be loaded on to textile materials with different fiber and fabric structures. The fibers can be degradable or nondegradable, and the fabric can be woven or nonwoven. In all these forms, drugs or other bioactive agents may be contained in the individual fibers or attached to the fabric surface. In recent years nonwoven fabrics produced by the electrospinning process have gained much attention for a variety of biomedical applications, including drug delivery.

Controllability is an important aspect of a delivery system. Preferably, a system should be controlled in such a way that the release of drugs is possible within a certain range, so that with any given system different release patterns should be feasible to respond to an individual patient's specific need. Material costs are also important. Commercially, there is no use in developing a drug delivery system in which the benefits are canceled out by the costs. To reduce costs, conventional production techniques are often beneficial in comparison to specially developed new techniques.

It should be pointed out that the choice of fiber-forming polymers is an important consideration during the preparation of drug-containing fibers. If a bioactive compound is to be incorporated within or inside the fiber, polymer selection is based upon the solubility of the drug in the polymer solution. An interesting development in this field is the synthesis of biodegradable polymers with a drug incorporated into the polymer backbone, whereby the drug is released upon degradation of the polymer. Special polymers have been developed that can degrade more easily in the presence of certain enzymes, such as enzymes found at infection spots, in which case drug release can be triggered by the onset of infection.

13.3 Fabrication of drug-releasing textile materials

Drug-releasing textile materials can be prepared by using a variety of drug-loading approaches such as coating, encapsulation, hollow fiber filling, ion exchange, complex formation, and direct conjugation methods. The following sections discuss these methods in detail.

13.3.1 Coating

Coating is one of the simplest approaches: a drug can be directly loaded on to the fabric surface by dipping the fabric in a drug-containing solution or coating it with drug-encapsulated microparticles or nanoparticles. In this method, the drug-loading efficiency depends on the nature of the fiber material as well as the bioactive agent or drug. For example, drugs with a high affinity toward a polymer will readily form a thin layer on the polymer surface. In many cases, drug-coated fabrics release a significant amount of the drug immediately following in vivo implantation. This inherent drawback makes coating less popular and unsuitable for releasing drugs over an extended period of time.

13.3.2 Encapsulation

A high drug loading within the fiber structure can be achieved by soaking the textile fabric in a drug-containing solution, allowing the drug to enter the fiber structure via a diffusion process. This encapsulation process is applicable to fabrics that can swell in drug-containing solutions. Drug loading and its subsequent release are solely diffusion controlled and dependent on the solubility of the drug and swelling of the fiber. Alternatively, the drug can be dissolved together with the fiber-forming polymer in a homogeneous polymer-drug solution before being spun into drug-containing fibers. This approach allows the required amount of drug loading and uniform distribution of the drug throughout the fiber structure.

13.3.3 Fiber spinning

Drug-containing fibers can be spun in a number of methods. Because of the high processing temperature, melt spinning is usually unsuitable for the production of drug-loaded fibers. In the wet-spinning process, the polymer-drug solution passes through the spinneret holes into a coagulation bath before being stretched, washed, and dried to form drug-loaded continuous fibers. In the electrospinning process, the drug-containing solution can be made into extremely fine diameters to produce nanofibers or microfibers for a variety of biomedical applications, including tissue engineering. Figure 13.1 shows the distribution of drug-containing particles in a textile fiber.

Because of their large surface areas, nanofibers are particularly suited for the development of drug-loaded fibers and textile materials. Technically, four different methods have been developed to yield nanofibers with complex architectures: electrospinning, coelectrospinning, tubes by fiber templates (TUFT), and wetting-assisted templating (WASTE). Electrospinning techniques have been widely used to produce functionalized nanofibers with diameters down to a few nanometers. Such nanofibers have been used with great success in areas such as bone tissue engineering, wound healing, carrying functional materials such as enzymes, etc. The TUFT and WASTE processes allow the manufacture of core–shell fibers with a high structural complexity to enable

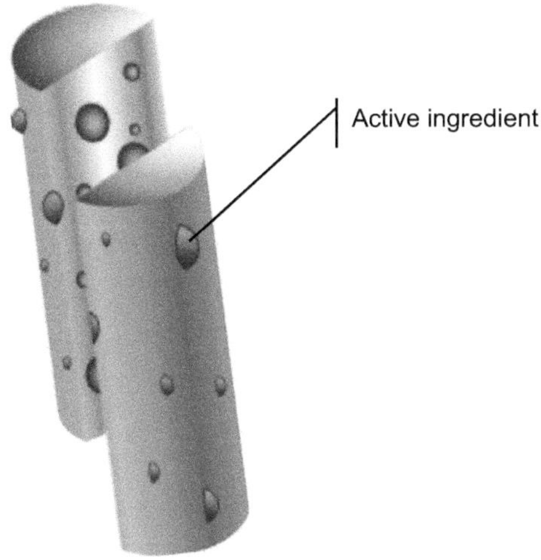

Figure 13.1 Distribution of drug-containing particles in a textile fiber.

further incorporation of functional nanoparticles. These fibers can serve as a highly versatile platform for a broad range of applications in widely different areas of medicine and pharmacy.

13.3.4 Hollow fibers

Hollow-fiber systems can be seen as small tubes which can be filled with a drug. The fiber wall usually consists of a permeable membrane that can be used in controlling drug release. Hollow fibers are excellent carriers for drugs because of the high surface area to volume ratio and the high loading flexibility, which can allow the filling of the hollow tubes with a liquid drug or a drug solution. Drug crystals dispersed in a polymer core can also be used. During the preparation process, the drugs are first mixed with a polymer material to produce a drug-loaded core granulate. This is then extruded in a coextrusion process to form the core of the hollow fiber, while another polymer is extruded to form the outer membrane of the fiber.

13.3.5 Bioconjugation

Bioconjugation techniques are used to attach bioactive molecules or drugs to the textile surface by chemical or physical means to achieve controlled release as well as site-specific delivery. In this method, the textile fabric must have readily available functional groups on its surface. Alternatively, functional groups can be created on fabric surfaces by using techniques such as plasma treatment, a wet chemical method, grafting, or cospinning. Fabric surface treatment with plasma can create a variety of functional groups depending on the gas used to create the plasma. These functional groups can then be used to conjugate with a variety of extracellular matrix components such as gelatin, collagen, laminin, and fibronectin to produce a highly biocompatible surface. Graft copolymerization is another attractive technique to conjugate bioactive molecules on the fabric surface, and is carried out through free radicals that are created by radiation or chemicals. For example, sulfonic groups can be introduced on to nonwoven polyethylene fabrics by radiation-induced graft polymerization of methacrylate and subsequent ring opening with sodium sulfite. These sulfonic functionalities can then be used to conjugate with silver ions in an effort to create antibacterial fabrics. Cospinning is another approach, where functional groups can be introduced on the fiber surface by combining different polymers with varied functionalities in their structure.

13.3.6 Ion complexes

Drugs can also be attached to fabric surfaces by utilizing the opposite charges present on the fiber surfaces and the drug molecules to produce ion complexes. In such complexes drug elution is controlled by the ability of the fabric surface to exchange preferentially with counter ions, where drug molecules are released in exchange for ions in the physiological system. Ion-exchange polymers and polymers containing ion-exchange groups are popular choices for such applications. Fabrics produced from such polymers are incubated in an ionic drug or bioactive molecule solution to produce ion complexes. For example, chitosan and alginate blends can be used to produce fibers with charged surfaces for forming complexes with charged drugs. A variety of bioactive molecules, such as dexamethasone, bovine serum albumin, growth factor, and avidin, can be mixed with either chitosan or alginate solution to produce drug-loaded fabrics. Positively charged molecules such as avidin are preferably incorporated into a polymer solution with an abundance of positive charges such as chitosan. The drug-chitosan solution can then be extruded into a solution containing alginate to produce drug-loaded polyion complex fibers. In contrast, mixing positively charged drug molecules with negatively charged polymers will result in aggregates that cannot be processed into fibers. Negatively charged molecules such as bovine serum albumin should be mixed with alginate and spun into a solution containing chitosan.

Ion-exchange fibers usually have a positive or negative electric charge that is compensated by mobile counter ions of opposite polarity. In cationic fibers, the bound ionic groups are formed by $-SO_3$, $-COO$, $-PO_3$, and $-AsO_3$, while in anionic fibers the bound groups are typically $-NH_3$, $-NH_2$, $-NH$, and $-S$. The principle of ion exchange is based on the electroneutrality requirement. The exchange is generally a diffusional process which is sensitive to concentration gradients, with the rate of ion exchange determined by the diffusion either within the exchanger itself or in the so-called diffusion boundary layer. Ion exchangers usually take up some counter ions in preference to others, and this can be used to control the release of active agents.

In ion complexes between drugs and ion-exchange fibers, the drugs act as the mobile counter ion. They can be released by ions in the body fluids, though this might disturb the homeostasis. Furthermore, in ex-vivo applications the amount of such ions will probably be too low to achieve sufficient drug release. Hence a co-solution of a salt, such as sodium chloride or sodium phosphate, is often used. The sodium ions will replace the drug ions and the latter will combine with the chloride or phosphate ions.

The salt can be encompassed in a gel based on gelatin, agar, or polyvinyl alcohol, which is then brought into contact with the textile fiber loaded with the drug. The salt will diffuse through the gel and fiber, and thus release the drug.

Ion-exchange fibers suitable for drug delivery are already commercially available. Examples are the Smopex® fibers, which consist of a polyethylene backbone grafted with other polymers such as polystyrene sulphonic acid (Smopex®-101), polyacrylic acid (Smopex®-102), or polyamide (Smopex®-108). Smopex®-301 is a styryl diphenylphosphine grafted polyolefin fiber which can be used to incorporate metal ions such as copper and zinc ions. In addition to commercially available ion-exchange fibers, it is possible to apply chemical modifications to natural fibers such as cotton, silk, and wool so that functional groups can be grafted to facilitate drug-loading capacities.

13.3.7 Nanoparticles

Textile materials can be treated with nanoparticles containing bioactive agents, with the nanoparticles permanently fixed on a fabric through chemical bonds, either by direct reaction of the side groups on the particle with the textile material or by means of a linker molecule. Nanoparticle encapsulation can circumvent possible alterations and loss of functionality of the bioactive agent, which could be the case in a direct reaction between the agent and the textile. However, linkers used for chemical attachment are difficult to engineer, and must be developed for the various agents individually.

In practice, the nanoparticles consist of a drug either surrounded by a polymeric shell or contained within a three-dimensional polymeric matrix, both micrometric to nanometric in size. Encapsulation can be achieved by several methods, such as interfacial polymerization, microemulsion polymerization, precipitation polymerization, and diffusion. In general, the drug is brought into contact with a set of monomers, oligomers, or polymers which assemble around the payload before polymerization is initiated to produce the final particles. An alternative way is to prepare the nanoparticles in the absence of the drugs, followed by absorption of the drugs by the nanoparticles. A wide range of applicable monomers, oligomers, and polymers is available for the preparation of nanoparticles, which often contain amine, hydroxyl, or sulfhydryl monomers or polymers combined with amine, hydroxyl, or sulfhydryl reactive monomers or polymers. Both hydrophobic and hydrophilic monomers can be used, with the choice depending on the drugs to be encapsulated.

Wool and other animal fibers contain hydroxyl, amine, carboxylate, and thiol groups that can easily react with nanoparticles. Synthetic fibers usually have a limited number of reactive groups, while cellulosic fibers have hydroxyl groups as the only reactives. To increase the possible particle-textile combinations, a bifunctional molecule can be used to link an amine-reactive particle with cellulose in order to attach reactive amine groups on the cellulose fibers.

For particle fixation, the textile material is exposed to a solution, dispersion, or emulsion of the textile-reactive particles by conventional methods such as soaking, dipping, spraying, fluid flow, or padding, in which a catalyst may be present in the medium. The actual fixation takes place in the curing stage, which can proceed while the textile is still in contact with the particle solution, or preferably after the drying stage. In the case of using a linker molecule, the textile material may first be treated with the linker and subsequently with the particle solution. The contact time can vary widely, from seconds to days, as can treatment temperatures and pressures.

13.3.8 Fullerenes

Fullerenes are an allotropic form of carbon, consisting of pure carbon atoms. A well-known example is C60, which is shaped like a football, so metal ions and other active agents can be enclosed in the inner cavity. Furthermore, fullerenes themselves can act as a guest molecule in complexes with macrocyclic ligands such as cyclodextrins.

The bound fullerenes are capable of reacting with other compounds such as aza-crown ethers, which in turn can bind other compounds. Because of their olefinic and electrophilic characteristics, fullerenes can react in several addition reactions, such as nucleophilic addition of amines, which can be used to achieve permanent fixation of fullerenes to polyamide and wool fibers.

13.4 Characterization of drug-loaded textile materials

The release of the drug is one of the most important characteristics of drug-loaded textile materials. Setting the release rate requires an understanding and careful manipulation of the release mechanisms, which include dissolution of the drug, desorption from the textile fibers, and diffusion toward the body fluid. The relative impact of each mechanism on the release rate depends highly on the chemical and physical properties of the fibrous material and the geometry of the product. By adjusting the extent of activity for one or more of the three mechanisms, the drug-release rate can be controlled to match an ideal profile for a specific requirement. For example, with drug-loaded wound dressings, the hydrophilicity and swelling of the fiber affect the rate of drug desorption and thus outward diffusion of drugs to the wound. Fiber diameter and geometry affect the diffusion rate within the fiber as well as the desorption rate on the surface. The type of material and the molecular orientation affect the biodegradation rate of the fiber and thus drug release via the degradation mechanism. By manipulating each of the three drug-release mechanisms, the rate of drug release from textile materials can be controlled.

The morphology and mechanical properties, degradation properties, drug-loading efficiency, and release properties can be tested by using a variety of analytical techniques both in vitro and in vivo. Some of these test procedures are described below.

13.4.1 Characterization of the base material

Drug-loaded medical textiles are composite materials in which the textile structure serves as the base material. Many traditional test procedures can be applied to the characterization of the base material. For example, the morphology of the fabric surface can be characterized by using microscopic techniques, including optical microscopy, scanning electron microscopy, and atomic force microscopy, which provide useful information about fiber diameter, deformities, and degradation properties. Mechanical testing of medicated textiles is highly desirable to determine the suitability of their application. Good mechanical stability is required to maintain the original shape and size of medicated textiles through repeated stretch and release cycles.

Thermogravimetric analysis and differential scanning calorimetric analysis can be used to determine the physical properties of both the base material and the encapsulated drug. The physical form of the encapsulated drug is an important factor in determining drug-release kinetics. For instance, highly crystalline drugs take longer to release than amorphous ones.

13.4.2 Characterization of drug-loading efficiency

The drug-loading efficiency can be determined by extracting the drug using solvents, followed by quantification tests. In vitro release of bioactive agents can be carried out in simulated body fluids (SBF) such as phosphate buffered saline (PBS) at 37 °C, and the released drug can be quantified using a suitable analytical method. In brief, known weights of medicated textiles are suspended in dissolution media such as PBS under constant agitation. Periodically, known quantities of dissolution media are withdrawn and analyzed for drug content using analytical techniques such as high-performance liquid chromatography, gel permeation chromatography, calorimetric, and mass spectroscopy methods. The choice of suitable analytical techniques for quantification tests depends on various factors, including the nature of drug, molecular weight, stability, convenience, and level of accuracy.

Biodegradability and biostability of drug-loaded textile materials are closely related to the drug-loading efficiency. Degradation studies can be carried out in SBF at a physiological temperature, where enzymes can be added to simulate in vivo conditions. During the test, a known weight of fabric is incubated in the degradation media and the changes in morphology, mechanical strength, and molecular weight are monitored at different time intervals to determine the extent of in vitro degradation.

13.4.3 Interface reactions

When loaded with drugs, textile materials can have irritating or sensitizing effects on the skin. With regard to the commonly used ingredients in the production of drug-loaded textiles, such as cyclodextrins and derivatives, fullerenes, and azacrown ethers, there is insufficient literature information about the biocompatibility of these materials. The safety data mark C60 as being irritating, hence the contact of fullerenes with skin and eyes should be avoided, although it is unclear how fixing and further complexation can alter its biocompatibility. With hollow fibers, irritation would stem from the choice of the wall material. By choosing fibers that are known to cause no bodily responses, the final product should be comfortable to wear without skin irritation. In addition, for textiles treated with nanoparticles, the textile materials themselves are unlikely to cause any undesired effects, as they are common materials which people are used to. For drug encapsulation, polymers with known and advantageous interactive properties should be used.

As far as interface reactions are concerned, the drug-loaded fibers will present no problems as long as the chosen polymers are considered to be safe, although the incorporation of drugs may have some effect on the properties of the base materials. One possibility would be the occurrence of surface modifications, which could be disadvantageous in respect of interface reactions.

The interaction between bioactive ingredients in the fiber and the skin surface is exemplified in the case of silver-containing wound dressings. As shown in Table 13.1, when a silver-containing wound dressing is applied on a wound surface, the interaction of silver ion in the dressing with the skin components is a complex process involving many factors which contribute to the release of silver ions into the wound. Ideally, once the silver ions are released from the dressing and consumed in their antimicrobial actions, they should be further transported by the metabolic processes and excreted out of the human body.

13.4.4 Controllability of drug release

Drug release from medical textile materials involves many factors, from the nature of the base material to the functional requirement of the drug. Drug release from cyclodextrin complexes is a typical case to illustrate these various factors, where the difficulties are associated with the fact that practically every cyclodextrin derivative as well as every cyclodextrin-drug combination has its own release pattern, due mainly to the variations of interactions governing the complexation strength. For weakly complexed drugs, dilution is a controlling factor as it shifts the equilibrium between absorption and desorption. For more strongly bound drugs in the drug-cyclodextrin complexes, competitive displacement and protein drug binding become important, with the former shifting the equilibrium by decreasing the available amount of free cyclodextrin, and the latter decreasing the available amount of free drugs. Drug release can be delayed by using pH-sensitive systems, for example, systems that are less soluble in water at low pH, but soluble in neutral or alkaline regions due to the ionization of an acidic

Table 13.1 The interaction of silver ions in a wound dressing with the skin

Level of penetration	Mode of interaction
Skin surface	Interaction with sweat, sebum, wound exudate, xenobiotic materials, bacteria/viruses, fungi
Wound/wound bed	Binds to protein and wound debris; absorbed by cells of wound margin
Dermal tissue	Interacts with connective tissues, fibroblasts, macrophages, growth factors, and cytokines; permeates blood vessels and granulation tissue
Hypodermal tissue	Systemic absorption by liver, spleen, bone marrow, and kidney urine

group. In oral administration, this means that drug release can be delayed until the complexes have passed through the stomach and reached the small intestine.

With ion-exchange fibers, the release pattern can be changed in several ways. Drug release increases with a decrease in electrolyte concentration, which can be explained by considering the drug equilibrium. It must be emphasized that the decreasing electrolyte concentration is a result of increased external volume. When the same amount of electrolyte is used, the external volume will be much higher in the case of lower concentrations, hence the equilibrium is shifted toward release. However, the decreased electrolyte concentration increases the electrostatic affinity between the drug and the fiber. An opposite situation is observed if the same external volume is used for different electrolyte concentrations, because in this case the shift in equilibrium would not be that great, and the affinity effect will be overriding. In general, with ion-exchange fibers, drug release can be controlled to some extent by choosing the appropriate fiber-drug combination and by changes in the electrolyte solution, concentration, composition, and pH.

For hollow fibers, the nature of the fiber membrane has a significant effect on the rate of drug release. In practice, the rate of drug release can be controlled through three mechanisms: diffusion, chemical reaction, and solvent activation. If the drug is present in a crystal state, the concentration of dissolved drug is determined by its saturation solubility, in which case the release rate can be controlled by the thickness and permeability of the fiber membrane. By choosing an appropriate membrane material and thickness, a specific release pattern can be established. When drug release is not diffusion controlled, changes in the release environment such as solvent composition can be used to control release.

With nanoparticles, the architecture of the particle shell or matrix can be formulated and fine-tuned to offer controlled release of its contents, ranging from constant but prolonged release to zero release. The particles can be formulated with desired characteristics via key structural features, such as cross-linking density, hydrophilic–hydrophobic balance of the copolymer units, and stiffness of the polymeric network. Furthermore, erodible or biodegradable particles can be used to combine the release mechanisms of diffusion and erosion. In addition, the particles can be designed to respond to different stimuli from the external environment by adjusting a number of parameters, such as pH, temperature, ionic strength, cosolvent composition, pressure, or electric field.

With fibers that incorporate drugs within the fiber structure, some control over drug release can be introduced by varying the following factors.

1. The type of polymer used in the fiber, which influences drug diffusion.
2. The diameter of the fiber, which controls diffusion distance.
3. The concentration of polymer used in the fiber, which influences fiber structure.
4. The amount of drug loaded in the fiber, which influences concentration gradients.
5. The biodegradability of the polymer, which enhances drug release through fast degradation.

Furthermore, the drug-loaded fibers can be combined with nonloaded fibers, either biodegradable or nonbiodegradable, which can be used to change the distance over which a drug has to diffuse before it is set free into the body or on to the skin.

13.5 Applications of drug-releasing textiles

Due to the high surface area and relatively simple incorporation processes, textile materials have been widely used for controlled drug delivery in many diverse applications, where medicated products are designed for wound care, transdermal drug delivery, antifungal and antimicrobial barriers, ophthalmic drug carriers, etc. Some of these applications are described below.

13.5.1 Drug-releasing textiles in wound care

The active ingredients used in wound management have evolved alongside the development of pharmaceutical agents and wound-management materials. Although topical pharmaceutical agents in the form of solutions, creams, and ointments have long been used for wound management, these traditional forms of delivery are inconvenient to use and it is difficult to achieve a sustained release profile. In recent years new generations of medicated dressings have been developed to incorporate novel chemicals with high therapeutic values, and to overcome some of the disadvantages associated with topical pharmaceutical agents. Instead of using traditional gauze and paraffin-impregnated gauze to deliver pharmaceutical agents, modern dressings use hydrocolloids, hydrogels, alginates, silicone gels, polyurethane foams, and films to carry bioactive ingredients. The incorporated drugs play important roles in the wound-healing process either directly or indirectly as cleansing or debriding agents for removing necrotic tissue, antimicrobial agents for the prevention or treatment of infection, and stimulating agents to aid tissue regeneration.

13.5.2 Antimicrobial textile products

The purpose of applying antibiotics and other antimicrobial agents is mainly to prevent or combat infections, especially for diabetic foot ulcers, burns, and surgical wounds where the incidence of infection can be high due to reduced resistance resulting from extreme trauma. Common antibiotics that are incorporated into dressings for delivery to wounds include dialkyl carbamoyl chloride, povidone-iodine, and silver, with silver in many forms widely used in advanced wound-management products. Other antibiotics delivered to wounds include gentamycin, ofloxacin, and minocycline. Freeze-dried fibrin discs have been used for the delivery of tetracycline. In general, the use of dressings to deliver antibiotics to wound sites can provide tissue compatibility, low occurrence of bacterial resistance, and reduced interference with wound healing. The use of lower antibiotic doses within the dressings also considerably reduces the risk of systemic toxicity. In addition, local delivery from dressings can overcome the problem of ineffective systemic antibiotic therapy resulting from poor blood circulation at the extremities in diabetic foot ulcers.

13.5.3 Medical textiles containing growth factors

Growth factors are involved in cell division, migration, and differentiation, protein expression, and enzyme production. The wound-healing properties of growth factors are mediated through the stimulation of angiogenesis and cellular proliferation, which affect both the production and the degradation of the extracellular matrix and also play a role in inflammation and fibroblast activity. Various growth factors are known to participate in the process of wound healing, including epidermal growth factor, platelet-derived growth factor, fibroblast growth factor, transforming growth factor, insulin-like growth factor, human growth hormone, and granulocyte-macrophage colony-stimulating factor. Most of these growth factors are recombinant proteins, and the choice of appropriate dressing is critical for effective release and action at the wound site.

13.5.4 Medical textiles loaded with nutritional supplements

Another group of active compounds that may be incorporated into medical textile materials is nutritional supplements such as vitamins A, C, and E. Vitamin A is involved in epithelial cell differentiation, collagen synthesis, and bone tissue development. It has also been shown to facilitate normal physiological wound healing as well as reversing the corticosteroid-induced inhibition of cutaneous wound healing and postoperative immune depression. Vitamin C is an essential compound for the synthesis of collagen and other organic components of the intracellular matrix of tissues such as bones, skin, and other connective tissues. It is also involved in the normal responses to physiological stressors such as accidents and surgical trauma, as the need for ascorbic acid increases during times of injury. In addition, vitamin C helps to improve immune function, particularly during infection. Vitamin E is capable of preserving important morphological and functional features of biological membranes, although its use in topical applications has been discouraged due to the problem of contact dermatitis.

In addition to the incorporation of vitamins, textile fibers can be used to carry zinc, copper, and other metal ions that are beneficial to the normal functions of the skin.

13.5.5 Production of cosmetotextile products

The ability of textile materials to carry bioactive ingredients can be used to micropackage cosmetic ingredients on to textiles. Microcapsules suitable for cosmetotextiles have mean diameters typically ranging from 1 to 10 μm. This compares to a range of 5–30 μm for a single textile fiber, from about 5 μm for microfibers to about 30 μm for coarse wool. Microcapsules can be modified in size, mechanical robustness, and permeability to customize the release profile to fit the intended functionality best.

According to the European Cosmetics Directive (76/768/EEC) Article 1: "A cosmetic product shall mean any substance or preparation intended to be placed in contact with the various external parts of the human body (epidermis, hair system, nails, lips and external genital organs) or with the teeth and the mucous membranes of the oral cavity with a view exclusively or mainly to cleaning them, changing their appearance and/or correcting body odors and/or protecting them or keeping them in good condition." Typical cosmetic functions are enhancing skin moisturization, firmness, or elasticity, UV protection, skin whitening or tanning, reduction of the outer appearance of cellulite, sebum regulation, and antiwrinkle, deodorant, and antiperspirant properties.

The cosmetic functionality defines the amount of a cosmetic product that is required to achieve efficacious performance. Skincare effects, such as moisturizing and tanning, require large amounts of the cosmetic product to be transferred to the skin in order to achieve perceivable effects, whereas fragrances only require the release of very small amounts to achieve effects. The efficacy of typical cosmetic moisturizing creams is usually demonstrated in clinical tests by applying 0.5–2 mg of the cream per cm^2 skin on to the tested area. An efficient cream should perform well at 0.5 mg/cm^2. If the concentration of the moisturizing ingredients in the cream is 20% (80% being water), this would correspond to a daily dose of 0.1 mg/cm^2. To achieve this dosage, a pantyhose covering an average of 1 m^2 of skin should contain 1000 mg of the cosmetic ingredients.

Since textile materials require frequent laundering during their life cycle, lipophilic cosmetic ingredients are often preferred, as it is relatively difficult to prevent water-soluble ingredients from being completely washed away during initial laundering. To maximize the efficacy of cosmetotextiles, the fabric composition and construction, garment design, and cosmetic finish need to work together to achieve the best effects. For example, cosmetotextiles with slimming effects will most likely work well if designed as a pair of elastic leggings with some degree of progressive nonmedical compression or as tight jeans, which allow close contact between the textile and the skin, and help to generate a massage effect that is beneficial to the transfer of an adequate amount of the cosmetic product from the fabric to the skin.

13.6 Summary

Textile materials are an important interface between human beings and the surrounding world. Due to their unique chemical and physical structures, textile fibers and fabrics are ideal carriers for drugs that can be directly transferred to the skin or other parts of the body. Because of the diverse ranges of fiber-forming polymers and fiber-making techniques, there is broad scope for the development of novel drug-containing textile materials for the improvement of human health.

Sources for further information

[1] Bartels VT, editor. Handbook of medical textiles. Cambridge: Woodhead Publishing Ltd; 2011.

[2] Boateng JS, Matthews KH, Stevens HNE, et al. Wound healing dressings and drug delivery systems: a review. J Pharm Sci 2008;97:2892–923.

[3] Robinson JR, Lee VHL. Controlled drug delivery: fundamentals and applications. New York: Marcel Dekker; 1987.

[4] Rouette HK. Encyclopedia of textile finishing. Berlin-Heidelberg: Springer-Verlag; 2001.

[5] Joshi M, Butola BS, Saha K. Advances in topical drug delivery system: micro to nanofibrous structures. J Nanosci Nanotechnol 2014;14(1):853–67.

[6] Stuart MA, Huck WT, Genzer J, et al. Emerging applications of stimuli-responsive polymer materials. Nat Mater 2010;9(2):101–13.
[7] Lu P, Ding B. Applications of electrospun fibers. Recent Pat Nanotechnol 2008;2(3): 169–82.
[8] Varshosaz J. The promise of chitosan microspheres in drug delivery systems. Expert Opin Drug Deliv 2007;4(3):263–73.
[9] Loftsson T, Brewster ME. Pharmaceutical applications of cyclodextrins, 1. Drug solubilization and stabilization. J Pharm Sci 1996;85:1017–25.
[10] Jaskari T, Vuorio M, Kontturi K, et al. Ion-exchange fibers and drugs: an equilibrium study. J Control Release 2001;70:219–29.
[11] Skundric P, Medovic A, Kostic M. Fibrous systems with programmed biological activity and their application in medical practice. Autex Res J 2002;2:78–84.
[12] Sawada Y, Ara M, Yotsuyanagi T, et al. Treatment of dermal depth burn wounds with an antimicrobial agent-releasing silicone gel sheet. Burns 1990;16:347–52.
[13] Maeda M, Kadota K, Kajihara M, et al. Sustained release of human growth hormone (hGH) from collagen film and evaluation of effect on wound healing in mice. J Control Release 2001;77:261–72.
[14] Gu F, Amsden B, Neufeld R. Sustained delivery of vascular endothelial growth factor with alginate beads. J Control Release 2004;96:463–72.
[15] Cangul IT, Gul NY, Topal A, et al. Evaluation of the effects of topical tripeptide-copper complex and zinc oxide on open-wound healing in rabbits. Vet Dermatol 2006;17:417–23.
[16] Lin SS, Ueng SW, Lee SS, et al. In vitro elution of antibiotic from antibiotic-impregnated biodegradable calcium alginate wound dressing. J Trauma 2000;47:136–41.
[17] Meaume S, Vallet D, Morere MN, et al. Evaluation of a silver releasing hydroalginate dressing in chronic wounds with signs of local infection. J Wound Care 2005;14:411–9.
[18] Aoyagi S, Onishi H, Machida Y. Novel chitosan wound dressing loaded with minocycline for the treatment of severe burn wounds. Int J Pharm 2007;330:138–45.
[19] Gombotz WR, Wee SF. Protein release from alginate matrices. Adv Drug Deliv Rev 1998;31:267–85.
[20] Peppas NA, Franson NM. The swelling interface number as a criterion for prediction of diffusional solute release mechanisms in swellable polymers. J Polym Sci Polym Phys Ed 1983;21:983–97.
[21] Ritger PL, Peppas NA. A simple equation for the description of solute release II. Fickian and anomalous release from swellable devices. J Control Release 1987;5:37–42.
[22] Siepmann J, Kranz H, Bodmeier R, et al. HPMC-matrices for controlled drug delivery: a new model combining diffusion, swelling and dissolution mechanisms and predicting the release kinetics. Pharm Res 1999;16:1748–56.

Biocompatibility testing for medical textile products

14

14.1 Introduction

Biocompatibility is defined as a measurement of how compatible a material or a device is with a biological system. With medical textile products, the purpose of performing biocompatibility testing is to determine their fitness for human use, and to see whether their use can have any potentially harmful physiological effects. In this respect, medical textile products share the same requirement for biocompatibility as medical devices.

The International Organization for Standardization (ISO) developed a series of standards for evaluating the biocompatibility of a medical device prior to a clinical study. In particular, ISO 10993-1:2009 clearly outlined its primary aim as the protection of humans from potential biological risks arising from the use of medical devices.

Biocompatibility testing usually begins with the extraction of leachable materials from the medical device or components at an elevated temperature, followed by analysis of the leachable extracts for potentially harmful chemicals or cytotoxicity. Once this in vitro testing has been completed, in vivo biological testing can be done based upon the device's intended use. The actual biocompatibility testing can range from a skin irritation test to hemocompatibility and implantation tests, with the turnaround time for tests ranging from 3 weeks to several months, depending on the requirement for specific test data. There are many available biocompatibility tests, including the following.

- Cytotoxicity tests, such as agarose overlay, minimum essential medium elution, direct contact, and MTT assay.
- Hemocompatibility tests, such as the Japanese Phamacopoeia (JP) hemolysis test and the American Society for Testing and Materials (ASTM) hemolysis test.
- Implantation testing, such as United States Pharmacopeia (USP) intramuscular implantation, ISO subcutaneous implantation, and ISO intramuscular implantation.
- Irritation and intracutaneous reactivity, such as USP/ISO intracutaneous reactivity, ISO mucosal irritation (vaginal, rectal, oral), ocular irritation test set by the ISO and the Organisation for Economic Co-operation and Development (OECD), intraocular irritation, and ISO/ Environmental Protection Agency (EPA)/OECD dermal irritation.
- Sensitization testing, such as ISO/EPA closed patch, dermal irritation, and the ISO maximization test.
- Systemic toxicity testing, such as the ISO acute systemic toxicity test, pyrogenicity test, USP/ Code of Federal Regulations (CFR) pyrogen test, ISO rabbit pyrogen test, USP systemic injection test, USP intracutaneous test, USP implant test, and USP safety test.

Medical Textile Materials. http://dx.doi.org/10.1016/B978-0-08-100618-4.00014-5

14.2 Regulatory requirements

Biocompatibility testing is a compulsory requirement to obtain regulatory approval for the marketing of a medical device. There are many regulatory authorities around the world which are responsible for establishing the detailed test requirements through national or regional legislations, such as the Food and Drug Administration (FDA) in the USA, the European Medicines Agency in the European Union (EU), the State Food and Drug Administration in China, the Pharmaceuticals and Medical Devices Agency in Japan, and the Medicines and Healthcare Products Regulatory Agency in the UK. In the EU the current medical device regulatory regime comprises three directives: the Medical Device Directive (MDD), the Active Implantable Medical Device Directive (AIMDD), and the In Vitro Diagnostic Medical Device Directive (IVDD). The MDD applies to any device used for "the diagnosis, prevention, monitoring, treatment or alleviation of disease" that does not act by pharmacological means. The AIMDD covers any device that is intended to be placed inside the body, remain after the procedure, and rely on a power source "other than that generated by the human body or gravity." The IVDD applies to any device intended "to be used in vitro for the examination of specimens derived from the human body." EU member states are required to implement the objectives of these directives in their own national laws.

To avoid the common mistakes that could slow a product's time to market, it is important for a medical device manufacturer to have a clear understanding of the regulatory requirement concerning its product. Regarding biocompatibility testing, there are many guidelines published by the various regulatory authorities. For example, the FDA published Blue Book Memorandum No. G95-1, which includes a part titled "Use of International Standard ISO 10993" under part one of "Biological Evaluation of Medical Devices." Based on these guidelines, product development engineers or regulatory affairs officers can design particular tests that will meet FDA and ISO requirements. In the USA these requirements were initially evaluated based on testing procedures specified by the USP. They continued to evolve to satisfy the testing requirements of as many countries as possible, resulting in the establishment of the ISO standards which aim to provide one set of standards that medical devices or materials must meet for worldwide approval. At this point, both the USP and the ASTM have generated equivalents of ISO 10993-1. One force driving this evolution is the Japanese Ministry of Health, Labor and Welfare. Early versions of ISO 10993 were relatively loose and allowed a lot of room for interpretation. However, the guidance documents for the Japanese market were very stringent, with no room for interpretation.

The current Japanese guidance document, titled "Test Methods for Biological Safety Evaluation of Medical Devices, Assessment of Medical Device, Notice 36, 2003," partially embraces ISO 10993, but it has technical specifications to perform tests that differ from the procedures specified in ISO 10993. As a consequence, many Japanese regulators do not readily accept testing conducted according to ISO 10993. Such reluctance will likely to continue until the technical sections of these two standards are harmonized. At present the technical conduct of many studies differs enough for the end point requirements of a Japanese test to be different from the requirements of ISO, and vice versa. Additional testing may thus be required to submit a medical product for Japanese approval.

In practice, the best course of action is to contract the biocompatibility testing to a contract research organization (CRO) that specializes in such testing and performs these tests according to good laboratory practices. This step appears to be the most prudent use of time and resources for a medical device manufacturer during the process of product registration. Prior to carrying out the tests, it is also important to get FDA agreement on the classification and testing scheme of the product, and a detailed description should be given to the CRO about the intended use of the product so that the technical experts can see clearly what tests they can offer to meet the requirements. A fully equipped CRO should have a standard battery of tests to meet ISO requirements as well as the requirements of other regulatory authorities.

Proper management of the relationship between a medical device manufacturer and a CRO is important in order to speed up completion of biocompatibility testing. Generally, the device manufacturer needs to consider three important questions.

1. How much sample is needed for the tests?
2. How much time will the tests require?
3. How much will the tests cost?

Similarly, an experienced CRO may ask questions such as the quantities of sample that the manufacturer should provide and the deadline for the test results before they are needed to file the submission. The required amount of samples will depend on the nature of the material or device in conformance to the relevant regulations, on which the CRO will normally advise. The guidelines for biocompatibility testing usually have strict requirements on test samples, to which the medical device manufacturer should adhere. Alteration of the sample requirements should be approved by the regulators, and the manufacturer is usually required to present justifications for doing so.

The actual biocompatibility testing requires careful planning in order to obtain the desired results within the shortest time. It should be pointed out here that the test systems for biocompatibility have a shelf life, in that the in vivo assays use animals within very specific weight ranges, while the in vitro assays use cells when they are at a certain confluence or density. To fit with the right phase of an animal's growth curve, the tests need to be timed to start when the animals are within the proper weight range and the cells are at the proper density. This choreography of growing cells or acquiring animals does not start until the CRO receives the test samples. There is usually a lag from the time when the samples are sent out to the time when the testing actually begins. This lag will vary from laboratory to laboratory and test to test. After the actual test, which should last a specified period of time, there may be additional evaluations, such as histopathology or clinical chemistry. All the data must then be evaluated and packaged by the study director in the form of a report.

14.3 Sampling requirements

Biocompatibility testing is usually performed on the finished product, and the test sample should go through any cleaning, processing, polishing, sterilization, etc. that the final product will go through. Sometimes the test sample would not lend itself to extraction because of its size or configuration. At other times the medical device manufacturer may prepare a scaled-down mock sample, in which case the mock sample should go through all the processes and have all the components in the proper proportions as the actual device. This may require retooling the whole production line, which may not be worth the potential material savings.

Careful consideration should be given to each test article to determine an appropriate approach to the testing. For example, a medical device may be a thin, sealed titanium can containing electronics. Extracting this device with the electronics contained within would not be appropriate, because there is little to no likelihood that a patient will be exposed to the electronics, and it would add mass to the test article and therefore increase the amount of extraction vehicle the can is exposed to. The result is a dilution of extractants leaching from the can, which can potentially mask toxic responses. In the case of hemodialyzers, the hollow fibers and housing are often tested separately, because of the different amount of extraction vehicle required.

For biocompatibility testing using cytotoxicity (ISO 10993-5), the samples are either tested directly, as in an agar overlay test, or are extracted, as in minimum essential medium elution. Extraction is a process in which the test material is typically subdivided, placed in an extraction vessel, and covered with the extraction vehicle. Polar and nonpolar extraction vehicles are used separately. Examples of polar extraction vehicles are water, cell culture media, and physiological saline, that is, 0.9 % aqueous NaCl solution, which is usually the preferred polar vehicle for biological assays. Examples of nonpolar vehicles include cotton seed oil and sesame seed oil.

The material requirements vary from test to test, particularly when carrying out direct-exposure tests. With extraction tests, 10 mL of extract are usually used per vehicle per extraction event. Exceptions to that general rule include the rabbit pyrogen test and repeat dose studies. The extraction ratios are dependent on the characteristics of the test sample. For example, the ratio for test articles with a thickness less than 0.5 mm is 6 cm^2 per 1 mL of vehicle, while test articles with a thickness of 0.5 mm or greater require 3 cm^2 per 1 mL of vehicle. Therefore, depending on the thickness, a 10 mL extract would require 60 or 30 cm^2 of test material. Surface area ratios should be the primary method of preparing extracts, and regulators around the world expect the submitters to provide testing that conforms to these ratios. If a fluid pathway or a clinically relevant extraction is used, it is then necessary to back calculate and show that the surface area to volume ratios met or exceeded the specified ISO ratios. Fluid pathway may meet or exceed the ratios, whereas clinically relevant ratios will likely fall short of the specified ratios, in which case the product will need to be retested.

To put the above principles into perspective, an empty manila folder, the cover of a paperback journal, and a staple from a standard desk stapler all have a thickness of about 0.5 mm. A typical small office sticky note has a thickness of about 0.1 mm and measures 3.5×4.8 cm, which is equivalent to a single-side surface area of 16.8 cm^2. When the surface area is calculated in the laboratory, all sides are calculated, and so the sticky note has a total surface area of 33.6 cm^2. Therefore, two sticky notes would be used to create 11.2 mL of extract.

Surface area to volume ratios are the preferred method of creating the extract; but when the geometry is complex, a weight to volume ratio of 0.2 g per 1 mL is used.

The definition of complex geometry is somewhat subjective, and the CRO may only consider tubings, slabs, cylinders, disks, and rings for using surface areas, while it is unlikely to calculate the surface area of a porous device or screw. In this case, the CRO will usually proceed with extracting at a weight to volume ratio unless specifically instructed by the sponsor to use other methods. Although the surface area can be calculated in computer-aided design drawings, it should still be noted whether a surface to volume ratio or a weight to volume ratio is used for products with complex geometry, so that the test results are not rejected and sent back by the regulatory authority due to incorrect or inconsistent extraction procedures.

When a device is complex and composed of various materials, the weight to volume ratio may be the more appropriate method to prepare the extract. For example, a simple three-material catheter may have the shaft made from one material and the hub made from another, plus the adhesive that holds the two together. All such materials are present in the finished device in differing thicknesses. Even though the surface areas may be easily calculable, a surface area to volume ratio could be inappropriate for such a case, since the shaft has far more surface area than the hub, while the hub is far thicker and there is far more material in the hub than the shaft. In addition, the adhesive is present at the interface between the shaft and the hub. In order to extract all the components at a consistent ratio, a weight to volume ratio should be used. In the case of large devices, it is also necessary to disassemble and sort the materials before the extract sample is prepared, with each material present in the same percentage as it exists in the finished device.

During the extraction test, the prepared samples are combined with the appropriate polar solvent in one vessel and the appropriate nonpolar solvent in another vessel. The vessels are then placed in an oven at 37°, 50°, or 70 °C for 72, 72, and 24 h, respectively, or an autoclave at 121 °C for 1 h. The temperature chosen should be the highest possible that does not degrade or deform the test article.

14.4 Description of some commonly required tests

14.4.1 Cytotoxicity testing

Cytotoxicity testing may be considered the canary in the coal mine of biocompatibility testing. Three basic categories of testing are required for every medical device, no matter how simple it may be: cytotoxicity, sensitization, and irritation and intracutaneous reactivity. The latter two are animal tests, and consequently are more expensive and take longer to get results. Therefore, when testing novel materials, cytotoxicity testing is typically performed first. This serves as a barometer as to how the rest of the testing is likely to proceed, and sponsors with questionable materials advance only after the results of cytotoxicity testing are known.

The purpose of cytotoxicity testing is to determine the biological reactivity of the mammalian cell lines following contact with the material or the extract of the material. During the test, cells are first grown on plates. Depending on the test being performed, the cells are either derived from the ovaries of Chinese hamsters or from fibroblasts of mice. Once the cells reach an appropriate confluence or density, the test article, or extract of the test article, is exposed to cell population. Following exposure, there is a period in which the plates will be observed for apparent cell death.

Cells are very sensitive and can be killed by minor changes in pH, salinity, or temperature. It is possible to have some incidence of this type of cell death without classifying the test article as cytotoxic. But if the vast majority of the cells are dead or dying, the test article will be classified as cytotoxic. Both in vitro and in vivo tests are required as part of a testing battery.

14.4.2 Sensitization testing

Sensitization testing establishes the potential of the test article to elicit an allergenic response. In particular, the testing is aimed at a delayed-type hypersensitivity (DTH) response. This is an immune response that takes a couple of days to develop. Poison ivy is a good example of a DTH response.

Two primary test methods are used for medical devices to satisfy the sensitization testing requirement. The first is the guinea pig maximization test (GPMT), and the second is the closed-patch test, also known as the Buehler. Owing to the use of an adjuvant to stimulate the immune system, the GPMT has been considered the more sensitive test. The GPMT has intradermal injections in the beginning of the study and topical applications at the conclusion. Due to the intradermal injections in the GPMT, the requirements specify that the test article be a liquid, suspendable powder, or extract. The Buehler is better performed with surface contact devices and chemicals.

The third commonly used sensitization test method is called the mouse local lymph node assay (LLNA). The LLNA is an acceptable test method, but device manufacturers have been using the guinea pig tests for decades and are satisfied with the results obtained using those methods. The LLNA, like any test, has its pros and cons. On the positive side, it is shorter in duration, uses animals of a lower phylum, and requires far less test material. However, it also generates radioactive waste and is not accurate in distinguishing a sensitization response from an irritation response, which increases the possibility of a false-positive outcome.

When proceeding with any sensitization study, it is necessary to know the irritancy potential of the test article. As mentioned, the LLNA may confuse irritancy with sensitization. In the two guinea pig tests, the animals are scored for sensitization by evaluating erythema and edema, that is, redness and swelling. Coincidentally, redness and swelling are also indicators of irritation. So it may appear that the GPMT and Buehler could give a false positive to an irritating test article; but the studies have safeguards to prevent this. They are designed with primary irritation animals that receive varying concentrations of the test article all on one day at the same time. Because the animals have not been exposed to the test article before this point, any redness or swelling is considered an irritation response and not an allergenic response. These results are then considered when preparing the dosing solutions for the main part of the study. At the challenge dose, which is the final determining dosing, the animals are dosed at the highest concentration determined to not produce any irritation. Any observation noted will be considered an allergenic response.

14.4.3 Irritation testing

Irritation testing determines whether the test article will cause irritation to the relevant tissue. Various models are available to test an article, the most common being the intracutaneous reactivity study (IC). In this test, extracts are typically used to inject five 0.2 mL boluses per extract into the skin of rabbits. The injections are usually the size of a mosquito bite. The skin sample area is observed every 24 h for 72 h following the injections and is scored for erythema and edema. The test sites are compared with control sites, and in order for the test article to pass, the difference of the average scores cannot exceed 1. The IC test can also be used with the test article directly if it is small enough to pass through a needle.

Other irritation models involve ocular, mucosal (oral, penile, bladder, rectal, and vaginal), and dermal tissues. The irritation test should use the most appropriate tissue for the particular device. For example, if the product is a contact lens, ocular irritation should be performed. A vaginal speculum should use a vaginal irritation study.

Even if appropriate tissue irritation is being performed, such as ocular irritation in the instance of the contact lens, consideration should be given to performing an IC test as well. On a topical dermal study, the test article is in contact with the epidermis. The epidermis is the first line of defense to protect the human body from the environment, and it has evolved to be fairly effective for that purpose. Therefore, some materials may not be detected as irritants by a topical dermal exposure, even if that is the most likely route of human exposure. Oral, rectal, vaginal, penile, bladder, and ocular tissues all have the ability to expel the test article or extract. The oral irritation study can utilize hamsters fitted with retaining collars to prevent a solid test article from being expelled.

14.4.4 Systemic toxicity testing

In discussing systemic toxicity, one must first understand that the term "systemic" refers to the way an animal is affected as a whole. In systemic toxicity testing the animal is exposed to the test article or its extract. Four categories of systemic toxicity testing exist. The first is called acute systemic toxicity, which is a single-exposure test with typically a 72 h observation. The second category is subacute toxicity testing, in which animals are exposed to the test article or extract for 14–28 days. In the other two categories, subchronic toxicity testing involves a 28–90 day exposure, while chronic toxicity testing requires a 6–12 month exposure.

In the toxicity studies described above, a shorter duration is indicated when the test is conducted using intravenous administration, while a longer duration is used for all other routes of administration. The category of testing required should be similar to the clinical exposure. Most medical devices will not need to pursue chronic toxicity testing. As the exposure period increases, so does the number of animals required. In continuous or repeat-dose studies, the animals are necropsied at the end of the study and pathological evaluation is performed on their organs. In addition, chemical, coagulation, and hematological evaluations are performed on the blood.

Pyrogenicity testing is also classified as a systemic toxicity test. It assesses whether or not a test article has the ability to cause a fever-like response. In the United States there are currently two types of tests for pyrogenicity, one in vitro and the other in vivo. The in vitro test, known as either the bacterial endotoxin test or Limulus amebocyte lysate (LAL), only detects pyrogens that are bacterial in origin, called endotoxins. The in vivo test is called the rabbit pyrogenicity test, and detects bacterial endotoxins as well as material-mediated pyrogens. Therefore, the rabbit pyrogenicity test is required during initial evaluation of a product, and the bacterial endotoxin test is used as a surveillance tool for bacterial contamination during manufacturing.

LAL uses a component of the blood of horseshoe crabs. In the presence of endotoxins, the LAL will clot. Using standard curves, the amount of endotoxin present is quantifiable. The rabbit pyrogenicity test involves monitoring the temperatures of rabbits for 3 h following an intravenous injection of the test article or the saline extract of the test article. If a rabbit's temperature increases 0.5 °C following the injection, the test is considered positive.

Pyrogenicity testing should not be confused with sterility testing. In the sterilization process, the bacteria on a material are killed but not removed. When the human body is exposed to these dead bacterial components, it recognizes the presence of bacteria and responds with a fever as if there were an active infection. In this sense, a sterile product is not necessarily nonpyrogenic.

14.4.5 Genotoxicity testing

Genotoxicity testing evaluates the test article's ability to cause damage to DNA, genes, and chromosomes. Because of this, genotoxicity testing is performed as a battery, which typically consists of a bacterial reverse mutation assay, also known as the Ames assay, a mouse micronucleus test, and a mouse lymphoma or chromosomal aberration test.

The Ames assay uses bacteria that have been mutated to require specific amino acids in order to survive. The bacteria are exposed to the test article in the presence of just enough of these essential amino acids to allow for a limited number of replications. A proliferation of bacterial DNA indicates that the bacteria reverse mutated back to wild-type bacteria and these limited amino acids were no longer essential to survive.

The mouse micronucleus test is an in vivo test in which mice are exposed to the test article or extract. Bone marrow is harvested from the animals and evaluated for the presence of micronuclei. Micronuclei are comprised of chromosomes or fragments of chromosomes, and are indicative of chromosomal damage.

The mouse lymphoma test uses a mutated mouse cancer cell line in which a partially damaged gene exists. When this gene is completely damaged, this mutated cell line is able to survive and replicate in the presence of a particular chemical. The cells are incubated within that chemical after exposure to the test article. If an increase in viability is detected, it would indicate that the test article was able to inactivate totally or damage the gene.

The chromosomal aberration study typically uses cells derived from Chinese hamster ovaries. These cells are encouraged to undergo mitosis, or cell division. They are then exposed to the test article and a chemical that stops the mitosis in metaphase stage. This is the stage in which all chromosomes are visible. At least 200 metaphase cells will be evaluated for visible damage to the chromosomes.

Genotoxicity testing is an important test which is required for anything that will come in contact with the body for over 30 days and or enter the body for more than 24 h. It is important because if something can cause genetic damage in these tests, it has the possibility to cause cancer.

14.4.6 Implantation testing

In implantation testing, the test material is placed into the bodies of animals. Various species can be used for the implantation test, ranging from mice to swine. Due to their physical sizes and the quantity of sites required to fulfill the guidelines, rabbits provide the most suitable model and are commonly used in implantation tests. Switching to rats is likely to increase the number of animals required to obtain the minimum amount of sites, which will increase the cost.

The tissue chosen for implantation should be the one that is most suitable for the test article. When in doubt, a muscle implant test can be performed first. In the muscle implant, the test article is contained in well-vascularized tissue, which affords the body the maximum opportunity to react with the test article. The best way to capture a potential reaction is to place small pieces of the test material into discrete pockets in the muscle along the spine of rabbits. That particular location is a single muscle mass, so there is minimal movement of the material after implantation.

Other sites for implantation include the subcutaneous space, bone, brain, and ocular, dental, or mucosal sites. If one of these locations is more appropriate for the finished test article, then that would be the better choice. Otherwise a muscle implantation test is generally used for biocompatibility evaluations.

The duration of these implantation studies is dictated by the clinical exposure time of the test article. Multiple time spans will need to be used. For a permanent implant, a minimum short term of 1–4 weeks and a long term of over 12 weeks should be performed, with a midterm of about 8 weeks. The point is to establish a steady state of the test site. If something from the test article leaches out, it would be noticed in the short-term implant but may have healed by 13 weeks. On the other hand, if something takes a while to start leaching or degrading, nothing may be observable at the 1–4 week time point but it will be evident at 13 weeks. Particular consideration should be given to time point selection when testing resorbable or biodegradable materials, based on the degradation rate of the test material.

At the end of the study, the implanted sites are retrieved and processed histopathologically. A pathologist evaluates the sites and scores them for a local tissue reaction. The scores from the control sites, which will be implanted with a proven biocompatible material, are subtracted from the scores of the test sites. The final score then categorizes the test material into a gradation of irritancy.

14.4.7 Hemocompatibility testing

Hemocompatibility testing includes a full battery of tests required of anything that comes into contact with blood. ISO 10993-4 clearly defines the categories of evaluations to be performed, as well as which types of medical devices need to have which tests performed. The five categories of hemocompatibility testing are thrombosis, coagulation, platelets, hematology, and the complement system. With the exception of thrombosis, all of these tests are in vitro assays.

In the thrombosis study, the test article is implanted into the vasculature of an animal. After a given period of time the implanted vessel is removed and the test article is observed for clot formations. The other four tests are carried out in test- tubes, with evaluations of specific actions to assess whether or not the test article has an effect on the blood's ability to clot or coagulate, can regulate an immune response, or can damage the cellular components of the blood.

14.4.8 Supplemental testing

Supplemental testing involves chronic toxicity testing and reproductive toxicity testing, as well as carcinogenicity studies and degradation studies. These are all long tests with high price tags. If such testing is required, a board-certified toxicologist or other professional with the proper education, training, and experience is needed to carry out the tests. These supplemental studies are not as clearly defined as the tests discussed earlier, and it would be advantageous to consult with someone who can put together appropriate testing for the product.

14.5 Summary

Biocompatibility testing is an essential requirement for regulatory approval of medical devices such as medical textile products, and it is important to follow strict testing protocol in order to avoid potential pitfalls that could delay a product launch. During the product development process, it would be sensible to start the process early enough to allow for thorough and complete testing. As with any business or scientific endeavor, it is important to communicate all details accurately and be as up to date as possible on all test requirements. It is often necessary to consult qualified professionals to provide the expertise that is required for the successful completion of the various tests.

Sources for further information

[1] ISO 10993–1:2009 Biological evaluation of medical devices part 1: evaluation and testing in the risk management process.
[2] ISO 10993–2:2006 Biological evaluation of medical devices part 2: animal welfare requirements.
[3] ISO 10993–3:2003 Biological evaluation of medical devices part 3: tests for genotoxicity, carcinogenicity and reproductive toxicity.

[4] ISO 10993–4:2002 Biological evaluation of medical devices part 4: selection of tests for interactions with blood.
[5] ISO 10993–5:2009 Biological evaluation of medical devices part 5: tests for in vitro cytotoxicity.
[6] ISO 10993–6:2007 Biological evaluation of medical devices part 6: tests for local effects after implantation.
[7] ISO 10993–7:2008 Biological evaluation of medical devices part 7: ethylene oxide sterilization residuals.
[8] ISO 10993–8:2001 Biological evaluation of medical devices part 8: selection of reference materials.
[9] ISO 10993–9:1999 Biological evaluation of medical devices part 9: framework for identification and quantification of potential degradation products.
[10] ISO 10993–10:2010 Biological evaluation of medical devices part 10: tests for irritation and delayed-type hypersensitivity.
[11] ISO 10993–11:2006 Biological evaluation of medical devices part 11: tests for systemic toxicity.
[12] ISO 10993–12:2012 Biological evaluation of medical devices part 12: sample preparation and reference materials.
[13] ISO 10993–13:1998 Biological evaluation of medical devices part 13: identification and quantification of degradation products from polymeric medical devices.
[14] ISO 10993–14:2001 Biological evaluation of medical devices part 14: identification and quantification of degradation products from ceramics.
[15] ISO 10993–15:2000 Biological evaluation of medical devices part 15: identification and quantification of degradation products from metals and alloys.
[16] ISO 10993–16:1997 Biological evaluation of medical devices part 16: toxicokinetic study design for degradation products and leachables.
[17] ISO 10993–17:2002 Biological evaluation of medical devices part 17: establishment of allowable limits for leachable substances.
[18] ISO 10993–18:2005 Biological evaluation of medical devices part 18: chemical characterization of materials.
[19] ISO 10993–19:2006 Biological evaluation of medical devices part 19: physico-chemical, morphological and topographical characterization of materials.
[20] ISO 10993–20:2006 Biological evaluation of medical devices part 20: principles and methods for immunotoxicology testing of medical devices.

Regulatory control of medical textile products

15

15.1 Introduction

Medical textile products are among a large group of products generally known as medical devices. The US Food, Drug & Cosmetic Act defines medical device as "an instrument, apparatus, implement, machine, contrivance, implant, in vitro reagent, or other similar or related article, including a component part, or accessory which is recognized in the official National Formulary, or the United States Pharmacopoeia, or any supplement to them, intended for use in the diagnosis of disease or other conditions, or in the cure, mitigation, treatment, or prevention of disease, in man or other animals, or intended to affect the structure or any function of the body of man or other animals, and which does not achieve any of its primary intended purposes through chemical action within or on the body of man or other animals and which is not dependent upon being metabolized for the achievement of any of its primary intended purposes."

Nowadays, medical devices are widely used in the healthcare industry, costing governments around the world a substantial amount of money. In 2000, the estimated 1.5 million different types of medical devices available on the market represented over US$ 145 billion. With innovation and the rapid advancement of technologies, medical devices are currently one of the fastest-growing industries. The global market value for 2006 alone exceeded US$ 260 billion.

In order to maintain safety and performance, governments around the world have put in place policies that address all elements related to medical devices, ranging from access to high-quality affordable products through to their safe and appropriate use and disposal. Although there is no single template that will respond to the needs of every country, regulations are required to ensure that the production facilities for medical devices meet the requirement for good manufacturing practice (GMP) and complex quality controls. Globally, governments are moving toward a harmonized regulatory system, because a proliferation of different national regulations increases costs, hinders access to healthcare technologies, and can even unwittingly jeopardize the safety of the patient.

In order for member states to adopt appropriate advanced regulatory systems, the World Health Organization is reinforcing its role in providing technical support to member states on areas such as vendor and device registration, training, and surveillance and information-exchange systems on the assessment of medical devices in use. The implementation of improved medical device regulatory systems can help protect the population from the risks of unsafe technology.

Medical Textile Materials. http://dx.doi.org/10.1016/B978-0-08-100618-4.00015-7

15.2 Medical device safety

Since all medical devices carry a certain degree of risk and could cause problems in specific circumstances, medical device safety can only be considered in relative terms. Many problems cannot be detected until extensive market experience is gained. For example, an implantable device may fail in a manner that was not predictable at the time of implantation, and the failure may reflect conditions unique to certain patients. For other devices, component failure can also be unpredictable or random. The current approach to device safety is to estimate the potential of a device becoming a hazard that could result in safety problems and harm. This estimate is often referred to as the risk assessment.

In the risk assessment for a medical device, a hazard is regarded as a potential for an adverse event and a source of danger. Risk is a measure of the combination of the hazard; the likelihood of occurrence of the adverse event; and the severity or overall impact. Risk assessment begins with risk analysis to identify all possible hazards, followed by risk evaluation to estimate the risk of each hazard. In general, risk assessment is based on experience, evidence, computation, or even guesswork. Risk assessment is complex, as it can be influenced by personal perception and other factors such as cultural background, economic conditions, and political climates. In practice, risk assessment of medical devices is based on the experience of healthcare professionals and on safety design engineering. In the United States, governmental risk assessment of medical devices is based mainly on recommendations from members of 16 medical specialty panels, and devices are categorized into three classes. In the European Union and Canada the classification schemes for medical devices are predominantly rules based. These rules categorize medical devices according to their perceived potential hazards. Canada assigns four classes of devices. The European Union assigns three classes, with Class II being subdivided into IIa and IIb, which effectively means four classes. The Global Harmonization Task Force (GHTF) is proposing a harmonized scheme for medical device classification.

In classifying medical devices, the potential areas of hazard that warrant consideration include the degree of invasiveness, duration of contact, the body system affected, and local versus systemic effects. An invasive device is usually considered to be a higher potential hazard than an equivalent noninvasive device. Similarly, devices that have a long duration of contact and affect vital organs such as the heart or the great arteries, or have systemic effects, are assigned higher classes of potential hazard or risk. The degree of regulation imposed on any device is proportional to its potential hazard. This approach is known as risk management. The first requirement of the "Essential principles of safety and performance of medical devices" recommended by the GHTF illustrates such an approach. It states that medical devices should be designed and manufactured in such a way that, when used under the conditions and for the purposes intended and, where applicable, by virtue of the technical knowledge, experience, education, or training of intended users, they will not compromise the clinical condition or the safety of patients, or the safety and health of users or, where applicable, other persons, provided that any risks which may be associated with their use constitute acceptable risks when weighed against the benefits to the patient and are compatible with a high level of protection of health and safety. This statement highlights the balance between risk and benefit of medical devices. The goal is therefore to maximize benefit and minimize risk.

The International Organization for Standardization (ISO) has produced a document (ISO 14971:2000) providing manufacturers with a framework including risk analysis, risk evaluation, and risk control for risk management in medical device design, development, and manufacturing, as well as for monitoring the safety and performance of the device after sale. Overall, optimum assurance of medical device safety has the following essential elements.

- Absolute safety cannot be guaranteed.
- It is a risk management issue.
- It is closely aligned with device effectiveness and performance.
- It must be considered throughout the lifespan of the device.
- It requires shared responsibility among the stakeholders.

15.3 Classification of medical devices

The term "medical device" covers a vast range of equipment and materials, from simple tongue depressors to hemodialysis machines. Like medicines and other health technologies, they are essential for patient care, for example, at the bedside, in a rural health clinic, or in large, specialized hospitals.

The GHTF has proposed a harmonized definition for medical devices, which means any instrument, apparatus, implement, machine, appliance, implant, in vitro reagent or calibrator, software, material, or other similar or related article intended by the manufacturer to be used, alone or in combination, for human beings for one or more of the specific purposes of:

- diagnosis, prevention, monitoring, treatment, or alleviation of disease;
- diagnosis, monitoring, treatment, alleviation of, or compensation for an injury;
- investigation, replacement, modification, or support of the anatomy or of a physiological process;
- supporting or sustaining life;
- control of conception;
- disinfection of medical devices;
- providing information for medical purposes by means of in vitro examination of specimens derived from the human body and which does not achieve its primary intended action in or on the human body by pharmacological, immunological, or metabolic means, but which may be assisted in its function by such means.

It is noted that an accessory is not considered to be a medical device. However, where an accessory is intended specifically by its manufacturer to be used together with a parent medical device to enable the device to achieve its intended purpose, it should be subjected to the same procedures and GHTF guidance documents as applied to the medical device itself.

It should also be noted that the definition of a device for in vitro examination includes, for example, reagents, calibrators, sample collection devices, control materials, and related instruments or apparatus. The information provided by such an in vitro diagnostic device may be for diagnostic, monitoring, or compatibility purposes. In some jurisdictions, reagents and the like may be covered by separate regulations.

Some products are considered to be medical devices in some jurisdictions, but there is not as yet a harmonized approach for these items. Examples include aids for disabled or handicapped people, devices for the treatment or diagnosis of diseases and injuries in animals, spare parts for medical devices, and devices incorporating animal and human tissues which may meet the requirements of the above definition but are subjected to different controls.

The GHTF classifies medical devices into the four classes.

Class I: generally regarded as low risk.
Class IIa: generally regarded as medium risk.
Class IIb: generally regarded as medium risk.
Class III: generally regarded as high risk.

Generally, Class III devices affect the functioning of vital organs and/or life-support systems, whereas Class I devices are noninvasive and do not come into contact or interact with the human body. Classifications are assigned by the risk that the medical device presents to the patient and the level of regulatory control which is needed to market the device legally.

In the FDA regulatory classifications there are three classes of medical devices: Class I, Class II, and Class III. This classification is based upon classifications for medical devices currently legally marketed in the United States. The FDA determines the device classification by the device's intended use and the risk that it presents to the patient. New medical devices are compared to legally marketed medical devices with the same intended use and technological characteristics to determine the classification.

In the FDA classification, Class I medical devices have the least amount of regulatory control, and present minimal potential harm to the user. Class I devices are typically simple in design and manufacture, and have a history of safe use. Examples include tongue depressors, arm slings, and handheld surgical instruments. Most Class I devices are exempt from premarket notification, and may be exempt from compliance with the GMP regulation.

Class II medical devices are those where general controls are not sufficient to assure safety and effectiveness, and existing methods/standards/guidance documents are available to provide assurances of safety and effectiveness. In addition to compliance with general controls, Class II devices are required to comply with special controls, which include:

- special labeling requirements;
- mandatory performance standards, both international and United States;
- postmarket surveillance;
- FDA medical-device-specific guidance.

Class II devices typically require premarket notification by submission and FDA review of a 510(k) clearance-to-market submission. A few Class II devices are exempt from the premarket notification. Information on Class II exempt devices is located within the device regulation 21 CFR 862–892. Examples of Class II devices include physiologic monitors, X-ray systems, gas analyzers, pumps, and surgical drapes.

Class III medical devices have the most stringent regulatory controls. For these medical devices, sufficient information is not available to assure safety and effectiveness through the application of general and special controls. Class III devices usually support or sustain human life, are of substantial importance in preventing impairment of human health, or present a potential unreasonable risk of illness or injury to the patient. Typically a premarket approval (PMA) submission to the FDA is required to allow marketing of a Class III medical device. A few Class III devices are required only to have a 510(k) cleared by the FDA to be marketed. Examples of Class III devices that require a PMA are replacement heart valves, silicone-gel-filled breast implants, and implanted cerebella stimulators.

15.4 Phases in the lifespan of a medical device

The major phases in the lifespan of a medical device include conception and development, manufacture, packaging and labeling, advertising, sale, use, and disposal, with each phase containing many different activities that may have an effect on the safety and performance of the device. Examples of how each phase can create health hazards are described below.

15.4.1 Conception and development

The scientific principles upon which a device is based are fundamental to its safety and performance. For example, a cardiac pacemaker should deliver a minute electrical impulse of a certain size and shape that simulates the natural functioning of the heart. Significant deviation from this may compromise safety and performance. Generally, a more complex device corresponds to a higher risk of user error, and soundness of concept and adequacy of design, construction, and testing, including verification, validation, and clinical trials, require scrutiny by scientific experts to ensure that design parameters and performance characteristics do not impose unwarranted risks.

15.4.2 Manufacture

Good functional medical devices are produced when the manufacturing process is adequately managed. However, poor manufacturing management can produce inconsistency in the quality of products, such that nonconforming products can filter through the production line to the market, even when the original prototype has been well designed. This consideration has led to the development of GMP for drugs, biological products, and medical devices. GMP is more commonly referred to as "quality systems in manufacturing."

15.4.3 Packaging and labeling

Properly packaged medical devices pose little risk to individuals handling them, even if the device is biohazardous. This highlights the importance of well-designed packaging systems in delivering clean, sterile, and protected medical devices to the point of use. Shipping is one of the hazards a medical device and its packaging must survive. Subtle damage can result during transportation and handling unless the total packaging system is designed robustly and can withstand various stresses. Well-sealed packaging is essential for medical devices that must be maintained sterile. Labeling is crucial in identifying the device and specifying instructions for its proper use. As for drugs, mislabeling can result in serious consequences for the user. Hazard warnings or cautions and clear instructions for use are very important.

15.4.4 Advertising

Advertising has the potential to create expectations and powerfully influence the belief in a medical device's capabilities. It is important, therefore, that medical device marketing and advertising are regulated to prevent misrepresentation of a device and its performance. Misleading or fraudulent advertising of medical devices may increase sales; but from the buyer's perspective, the purchase of an inappropriate medical device is a waste of money that may deprive the patient of more appropriate treatment and could lead to patient or user injury.

15.4.5 Sale

The sale of medical devices is a critical stage that leads to a device being put into actual use. If the vendor is not subjected to regulation, there is a high risk of exposing the public to low-quality or ineffective devices.

15.4.6 Use

Users of medical devices can have a profound effect on their safety and effective performance. Unfamiliarity with a certain technology or operating procedure, and the use of products for clinical indications outside the scope of those specified in the labeling, can cause device failure even in the absence of any inherent design or manufacturing defects. Within the clinical engineering community, it is widely believed that user error underlies at least half of all medical-device-related injuries and deaths. The reuse of disposable devices contrary to the manufacturer's instructions, and without proper control or precautions for minimizing associated risks, can be dangerous. The lack of appropriate calibration and maintenance of medical devices can seriously jeopardize their safety and performance.

15.4.7 Disposal

Disposal of certain types of devices should follow specific and stringent safety rules. For example, devices that are contaminated after use (e.g., syringes) or those that contain toxic chemicals can present hazards to people or the environment and must be disposed of properly.

15.5 The role of each participant in ensuring the safety of medical devices

In general, it is people who manage each phase in the lifespan of a medical device, and these people should be identified and called on to participate in ensuring device safety. Several different groups of people are involved in the commercial process of delivering a medical device to its final user. The role of each participant in ensuring product safety is described below.

15.5.1 The manufacturer

As the creator of the device, the manufacturer must ensure that it meets or exceeds the required standards of safety and performance during the phases of design, development, testing, manufacturing, packaging, and labeling. As pointed out earlier, "user error" is one of the most common sources of product malfunctions, and may result from a mismatch between variables—for example the operator, device, task, or environment. By incorporating human-factor engineering principles in design, and appropriate training for users, the risk of user errors can be minimized.

15.5.2 The vendor

The vendor provides the interface between product and user, and should ensure that the products sold comply with regulatory requirements. With increasing public interest in health and a competitive market place, vendors should be careful to avoid making misleading or fraudulent claims about their products or issuing false compliance certificates. In addition, used or refurbished devices should be clearly labeled as such. Since medical devices often require specialized training from the manufacturer for proper use and service, the vendor should make training a condition to the manufacturer or importer in agreeing to sell the device. In turn, vendors should take responsibility in supporting or training their customers. Participating in postmarket surveillance, such as receiving and reporting customer complaints and incident reports, is critical for ensuring medical device safety and performance. The vendor must fulfill those obligations specified by the regulatory authority. For example, the vendor must make arrangements for processing complaint and incident reports relating to device safety and performance.

15.5.3 The user

The user should have qualifications and training in the proper use of a device, and should be familiar with the indications, contraindications, and operating procedures recommended by the manufacturer. It is crucial that experience gained with medical devices be shared with other users, the vendor, and the manufacturer to prevent future problems. This can be done by reporting any incidents to a coordinating center from which warnings can be issued.

When using medical devices, users should always bear in mind that the safety and health of patients are in their hands. The user has the responsibility to employ a medical device only for the intended purposes. The user also has the responsibility to ensure proper maintenance of medical devices during active use and safe disposal of obsolete devices.

15.5.4 The public

The public are the ultimate beneficiary of medical devices. They should be fully aware that all devices carry a certain risk, and that they can help to promote safety and performance through self-education and by putting customer pressure on manufacturers to comply with standards. Medical devices are increasingly available for home use, making the public the direct users. Purchasers of home-use medical devices should be aware of associated risks and take the responsibility to become educated in the functions and correct operating procedures for those devices.

15.5.5 The government

The government has the responsibility to oversee the efforts of manufacturers and vendors and ensure that medical devices sold or made available in the country are safe and effective. It should provide leadership in creating healthy cooperation among stakeholders in establishing policies and regulations that are fair and clear to all. Policies and regulations should be reviewed periodically to respond to changes in technologies by incorporating appropriate amendments.

In summary, shared responsibilities by all stakeholders are required to ensure the safety and performance of medical devices.

15.6 Governmental regulation of medical devices

To protect the health and safety of its people, a government fulfills its duties through the establishment and implementation of regulations to govern the two critical elements affecting safety and performance: the product and its use. Premarket review contributes to product control, and postmarket surveillance ensures that medical devices in use continue to be safe and effective. There is an important third element, which is the representation of the product to the user. This is controlled through labeling during the premarket stage and advertising during the sale of the product. In general, premarket control is performed to ensure that the product to be placed on the market complies with regulatory requirements. Labeling and advertising control is maintained for correct product representation. Placing-on-market control ensures establishment registration, product listing, and after-sale obligations. Postmarket surveillance and vigilance ensure the continued safety and performance of devices in use.

15.6.1 Product control

Although different authorities have different systems of premarket review, they all apply a similar risk-management philosophy in which all medical devices must satisfy safety and performance, quality system, and labeling requirements. The degree of regulatory scrutiny increases with the potential risks of the medical device, as evidenced by the risk-based device classification system proposed by the GHTF.

Governments around the world approve the sale of medical devices in many different ways. In Australia the Therapeutic Goods Administration issues an Australian Register of Therapeutic Goods number to devices cleared for the market. In Canada a device licence is awarded by the Therapeutic Products Directorate. In the European Union, after receiving an EC certificate from a notified body, the manufacturer places the CE mark on the device. In Japan a shounin is issued by the Pharmaceutical and Medical Safety Bureau of the Ministry of Health, Labor and Welfare. In the United States the manufacturer of the device receives a marketing clearance (510(k)) or an approval letter (PMA) from the FDA.

In the European system, manufacturers of devices of Classes II and III, as well as Class I devices with either measuring function or sterility requirements, must submit to the regulatory authority a declaration of conformity to the appropriate EC directives, and details of the conformity assessment procedure. In addition, for higher-risk-class devices that require design examination or type examination, the corresponding EC certificates issued by a notified body must be submitted to the competent authority. Other Class I medical devices are exempt from premarket submissions, although they must follow the essential principles of safety and performance in their design, construction, and labeling requirements.

In the United States, most Class III and new devices that are not substantially equivalent to a legally marketed product that does not need a PMA application require clearance through the PMA or Product Development Protocol processes. Most Class II and some Class I devices require a premarket entry notification known as the 510(k), which is essentially an information package for the FDA that receives less stringent review than the PMA process. The 510(k) submission must demonstrate how the proposed medical device is substantially equivalent to a medical device that is already on the US market. Most Class I and some Class II (low-risk) devices are exempt from 510(k) submission before sale, but are still subject to general control requirements.

15.6.2 Vendor establishment control

Vendor information facilitates governments in tracking medical device vendors. In Australia the sponsor must hold an enterprise identification number before being permitted to apply to register or list products. Similarly, in Canada any individual or company wishing to sell medical devices must apply for permission to obtain an establishment licence. The European Union requires a responsible person of the vendor establishment with a physical address in Europe be registered. In Japan, medical device sales organizations must have a licence called "hanbai-gyoo" or "hanbai todoke." In addition, importers are required to have a licence called "yunyu hanbai-gyo." In the United States, the establishment (manufacturer, initial importer, specifications developer, contract sterilizer, repackager, and/or relabeler) must be registered with the FDA.

15.6.3 Postmarket surveillance and vigilance

It is critically important that the safety and performance of medical devices are continually assessed when they are in use, as these characteristics can only be proven if one measures how a device stands up in these conditions. No amount of rigor in the premarketing review process can predict all possible device failures or incidents arising from misuse. It is in actual use that unforeseen problems related to safety and performance can occur. It should be pointed out here that different terms in medical device vigilance and postmarket surveillance have varying meanings in different countries. In general, postmarket surveillance is a broad term that covers all monitoring activities of medical devices in use. The two principal activities within surveillance are "postmarket surveillance studies" and "adverse event reporting."

In postmarket surveillance studies, specific and structured data collections are required of the manufacturer in one of two situations: as a condition of product approval, or to reaffirm product safety when postmarket adverse event reports suggest that premarket safety claims are inconsistent with actual use and result in unacceptable risk.

Japanese authorities and the FDA actively make use of surveillance data collection to augment the findings of premarket trials. Adverse event reporting requires the registration and investigation of adverse events relating to the use of a device, and the relevant authority should oblige the manufacturer to recall or modify a defective device. All founding members of the GHTF have mandatory requirements for vendors and manufacturers to report all device-related events that have resulted, or could result, in serious injury or death. In some countries, mandatory adverse event reporting is also extended to users. In addition, postmarket surveillance is interrelated to the quality system requirements described in the following section.

15.7 Quality system requirements

A quality system is defined as the organizational structure, responsibilities, procedures, processes, and resources needed to implement quality management. Quality system standards are "generic management standards." The international quality system standards for medical devices, ISO 13485:1996 and ISO 13488:1996, are issued by the ISO. ISO 13485:1996 includes all the elements of ISO 9001:1994 plus a set of minimum supplementary requirements for medical devices. ISO 13488:1996 specifies the quality system requirements for production and, when relevant, installation and servicing of medical devices in conjunction with the application of ISO 9002. In general, regulations for quality systems may cover the methods, facilities, and controls used by a manufacturer in the design, manufacture, packaging, labeling, storage, installation, servicing, and postmarket handling of medical devices. In this way, quality system requirements can influence all phases in the lifespan of medical devices.

In practice, the applicable requirements depend upon the risk class of the device and the regulatory system of the country. Design control is normally not required for regulatory scrutiny in medium- to low-risk devices. In the manufacturing process, quality system requirements impose strict quality assurance on every aspect of production. The result is a tightly controlled manufacturing system, commonly known as GMP, which reduces the likelihood of nonconforming products. This practice ensures consistency in quality and provides the basis for greater reliability in device safety and performance.

Elements of the quality system are periodically subject to audits, management review, and corrective or preventive actions that will maintain product quality. Continuous monitoring and corrective action requirements are interrelated to the postmarket surveillance previously described. The key advantage regarding quality systems is that they represent a preventive approach to assuring medical device quality, versus the previous reactive approach by inspection and rejection at the end of the manufacturing line. Prevention has been proven to be more efficient and cost effective in controlling manufacturing processes and maintaining medical device quality.

15.8 FDA regulation of medical devices

In the FDA, the Center for Devices and Radiological Health is the division responsible for medical device regulation. Many of the regulations enforced by the FDA with regard to medical devices can be found in the Code of Federal Regulations, Title 21, Food and Drugs, Part 500 to Part 1299, which is abbreviated to 21 CFR 800 to 1299. The regulation requires clearance by the FDA before medical devices can be sold in the US market. In general, medical devices are subject to general controls and premarketing and postmarketing regulatory controls.

15.8.1 General controls

General controls include the following.

1. Establishment registration by manufacturers, distributors, repackagers, and relabelers.
2. Listing with the FDA of medical devices to be marketed.
3. Manufacturing of devices in accordance with GMP.
4. Labeling medical devices in accordance with labeling regulations 21 CFR 801 or 21 CFR 809.
5. Reporting of adverse events as identified by the user, manufacturer, and/or distributor of the medial device.

15.8.2 Premarketing controls

Premarketing controls are specific to devices and device classifications. Premarketing controls for a medical device may include clearance to market by a 510(k) or approval to market by a PMA.

The majority of Class II medical devices are cleared to market by submission and FDA review of a 510(k). The 510(k), also known as a premarket notification submission, identifies characteristics of the new or modified medical device as compared to a device with similar intended use currently legally marketed in the United States. The currently legally marketed device is referred to as the "predicate" device. To obtain this clearance, companies must file an application and exhibits to demonstrate that the new device is substantially equivalent to a device already legally marketed, including substantial scientific and technical information. The information required in a 510(k) submission is defined in 21 CFR 800 to 1299, and includes the following.

1. Device trade or proprietary name, common or usual name or classification, class of the device (Class I, II, III).
2. Submitter's name and address, contact person, telephone and fax numbers, representative/consultant if applicable.
3. Name and address of manufacturing/packaging/sterilization facilities, registration number of each manufacturing facility.
4. Action taken to comply with the requirements of the special controls.
5. Proposed labels, labeling, and advertisements to describe the device, its intended use, and the directions for its use.
6. 510(k) summary or a 510(k) statement.
7. For Class III medical device, a Class III certification and a Class III summary.
8. Photographs of the device, engineering drawings of the device.
9. Identification of the marketed device to which equivalence is claimed, including labeling and description of the medical device.
10. Statement of similarities and/or differences with the marketed device.
11. Data to show consequences and effects of a modified device, performance data (bench, animal, clinical).
12. Sterilization information (as applicable).
13. Software development, verification, and validation information.
14. Hardware design and development information.
15. Information requested in specific guidance documents (as applicable).
16. Kit certification statement (for a 510(k) submission with kit components only).
17. Truthful and accurate statement.

Depending on the complexity of the new or modified medical device, the FDA review of a 510(k) submission takes from 20 to 90+ days. The review process lasts longer if complex changes or comparisons are required to support the safety and effectiveness of the new or modified device.

A PMA is required for FDA approval of medical devices that present significant risk to the patient and/or require significant scientific review of the safety and effectiveness of the device prior to commercial introduction. Most Class III medical devices require a PMA. Section 515(c) of the Federal Food, Drug, and Cosmetic Act specifies the required contents of a PMA, which include the following.

- Full reports of all information, published or known to or which should reasonably be known to the applicant, concerning investigations which have been made to show whether or not such device is safe and effective.
- A full statement of the components, ingredients, and properties and of the principle or principles of operation of such device.

- A full description of the methods used in and the facilities and controls used for the manufacture, processing, and, when relevant, packing and installation of such device.
- An identifying reference to any performance standard under Section 514 which would be applicable to any aspect of such device if it were a Class II device, and either adequate information to show that such aspect of such device fully meets such performance standard or adequate information to justify any deviation from such standards.
- Such samples of such device and of components thereof as the FDA may reasonably require, except that where the submission of such samples is impracticable or unduly burdensome, the requirement of this subparagraph may be met by the submission of complete information concerning the location of one or more such devices readily available for examination and testing.
- Specimens of the labeling proposed to be used for such device.
- Such other information relevant to the subject matter of the application as the FDA, with the concurrence of the appropriate panel under Section 513, may require.

The FDA has a statutory 180-day review cycle for PMA applications. Often PMA applications require medical advisory board review prior to the FDA granting approval to market the device. A facility inspection verifying the systems for the manufacture of the medical device is usually performed prior to PMA approval. FDA approval of a PMA often takes significantly more than 180 days.

15.8.3 Postmarketing controls

Postmarketing controls include establishment registration, device listing, quality system compliance inspection, and medical device reporting (MDR).

Establishments involved in the production and distribution of medical devices intended for marketing in the United States are required to register with the FDA. The establishment registration requirements are defined in 21 CFR 807. Foreign establishments engaged in the manufacture, preparation, propagation, compounding, or processing of medical devices that are imported, or offered for import, into the United States must register their establishments and provide the FDA with the name of the US agent representing them. They must also provide the FDA with a list of the devices that they are exporting to the United States.

Most medical device establishments required to register with the FDA must list the devices they have in commercial distribution. Medical device listing is completed by the classification name the FDA has assigned to the medical device illustrated in 21 CFR Parts 862–892. Device listing should be completed within 30 days of entering a device into commercial distribution in the United States. Device listing is updated when a new classification of device is entered into commercial distribution, the marketing of a device classification is discontinued, or marketing of a discontinued device is started again.

Medical device establishments are required to manufacture their products under the GMP requirements defined in 21 CFR 820. GMP is also referred to as the quality system (QS) regulation. A QS is required for the design, manufacture, packaging, labeling, storage, installation, and servicing of finished medical devices intended for commercial distribution in the United States. The QS regulation is similar to ISO

9001:1994, with particular requirements from the FDA. The regulation covers quality management and organization, device design, buildings, equipment, purchase and handling of components, production and process controls, packaging and labeling control, device evaluation, distribution, installation, complaint handling, servicing, and records. The FDA may audit a manufacturing facility for GMP compliance every 2 years. The FDA compliance inspections are based upon results of prior audits, device risks, recall of devices, and FDA initiatives in regard to particular classifications of devices.

The MDR regulation requires firms which have received complaints of device malfunctions, serious injuries, or deaths associated with medical devices to notify the FDA of the incident. The requirements for MDR are defined in 21 CFR 803, and include the following.

1. A written MDR procedure.
2. MDR event files.
3. Individual adverse event reports.
4. Five-day MDR reports of remedial action taken to prevent unreasonable risk of substantial harm to public health.
5. Thirty-day MDR reports of deaths, serious injuries, and malfunction attributed to the medical device.
6. Baseline reports for the medical device.

15.9 Summary

Medical textile products are a special type of medical device with strict safety and functional requirements. To ensure minimum levels of malfunction during their clinical application, considerable efforts need to be taken during the various phases of the product lifecycle, from conception to development, manufacture, packaging and labeling, advertising, sale, end use, and disposal. Each participant in this chain shares the responsibility for ensuring products meet safety and functional requirements, which are outlined in many related government regulations.

Sources for further information

[1] World Health Organization. Medical device regulations: global overview and guiding principles. Geneva: WHO Library Cataloguing-in-Publication Data; 2003.

Research and development strategy for medical textile products

16

16.1 Introduction

Product development is about developing technical solutions to meet market needs. In the medical textile industry, market needs vary widely—from wound dressings to surgical sutures, protective garments, absorbent pads, vascular prostheses, and many others. At the same time, a large variety of technologies are used to make innovative products. Many types of organizations work together in the research, development, manufacture, and sale of medical textile products that are used by patients with many different clinical requirements. Figure 16.1 outlines the industry structure for medical textile products.

Similar to many other industries, the medical textile industry is made up of a large number of companies, each engaged in a diverse range of business activities. To understand the nature of product development in this particular field, it is useful to have an overall understanding of the various players involved. According to the five-forces model proposed by Michael Porter of Harvard Business School, the five types of players in the medical textile industry are as follows.

1. Buyers: In the medical textile industry, a buyer can be a large multinational corporation that purchases products from small high-tech companies with specialist technologies and then distributes them to end users, or hospitals that buy products from various types of producers.
2. Suppliers: A medical textile company needs many types of suppliers. Apart from the usual business suppliers such as accountants and legal firms, the main suppliers to medical textile companies are raw material and technical service providers. Research contractors are another type of suppliers, as are suppliers of components and equipment.
3. Industry rivals: In the medical textile industry, many firms are engaged in the same or similar product areas. These rivals compete with each other on the quality and price of their products. For a company to have competitive advantages over its rivals, it needs to develop cutting-edge technologies and manufacturing capabilities so that its products can have superior performance but be manufactured at a low cost.
4. Substitutes: Many types of medical textile products have similar properties, used in a similar way for the same end uses. These so-called substitute products compete with each other on quality and price.
5. New entrants: In any industry that is highly profitable, more investment is attracted because of the high return on investment. To avoid the pitfalls of overcompetition, medical textile companies should prevent new rivals from entering into their business areas by creating barriers to entry, such as by applying patent protection or developing novel manufacturing technologies.

Medical Textile Materials. http://dx.doi.org/10.1016/B978-0-08-100618-4.00016-9

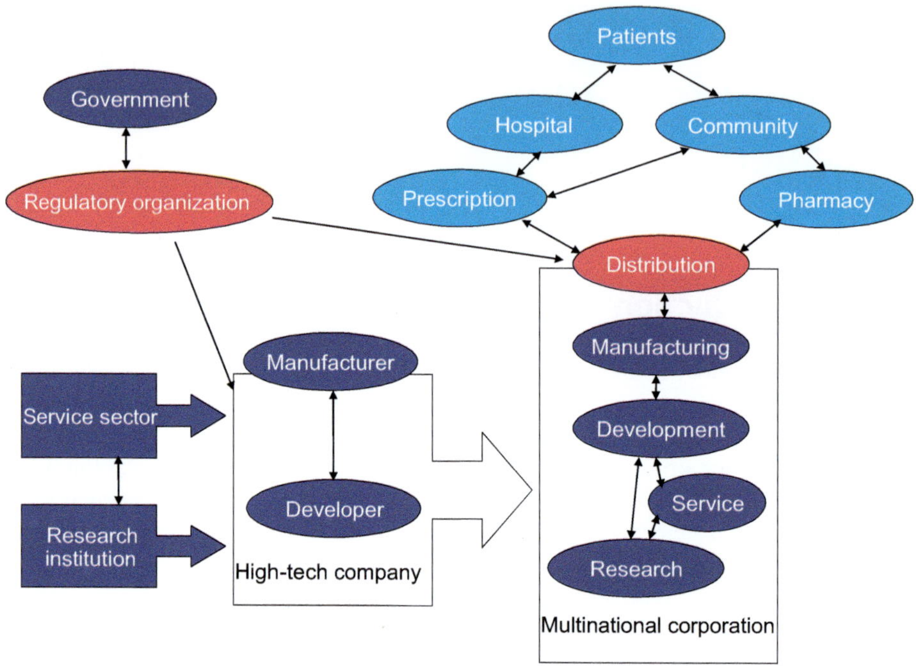

Figure 16.1 A schematic illustration of the industry structure for medical textile products.

In addition to these five types of players, industrial regulators are becoming increasingly important in many industries. This requires serious managerial attention, since medical textile products cannot be legally sold without regulatory approval. Industrial regulators have become a powerful force in the medical textile industry through their roles in establishing product standards, granting licenses, implementing environmental and safety rules, monitoring quality systems, and regulating many other critical issues related to the statutory requirements on maintaining the safety and performance of medical textile products.

16.2 The product development process

Figure 16.2 shows the product development process from the beginning, when an idea is formed, to the point when customers can use the product. The product development cycle normally begins with an idea to serve a particular customer need, followed by a series of research, development, manufacturing, marketing, distribution, customer service, and other related commercial activities that comprise a full product value chain. Because of the various types of risks involved, the success of medical textile product development depends on the careful execution of the many tasks that are needed to complete the development process. It has been reported that on average, for every 100 projects that go into development, 63 get killed before the final stage, 12 are

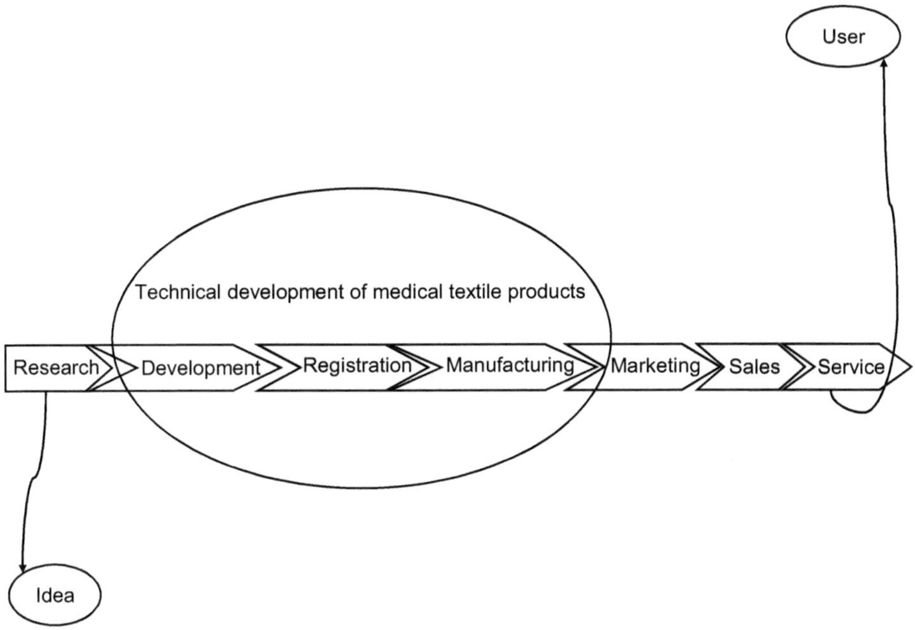

Figure 16.2 The product development process.

commercial failures, and only 25 become commercial successes. Therefore, during the development of medical textile products, the need to develop a truly innovative product should always be balanced with the costs associated with failures at the later stages of the development cycle.

A staged new product development (NPD) process with go/kill points is a useful way to control the inputs and outputs of the many design activities. In a five-stage NPD process, the objectives of each stage are set and pass/fail criteria are defined before the detailed development work proceeds. A documented stage review is held at the end of each phase of the development process. A product development project may proceed or terminate according to the results of each stage review. By applying strict reviews at the end of each stage, the staged NPD process ensures that only the truly successful project can proceed to full-scale manufacturing, by which time a large amount of capital will have been invested.

In staged NPD process, product development can be divided into the following five stages.

1. Concept formation.
2. Initial exploration.
3. Development.
4. Product realization.
5. Product launch.

Various issues are addressed during each stage. A project manager is responsible for ensuring that, at the end of each stage, reviews are held at appropriate times with appropriate personnel. The outcome of these reviews should be a decision to abandon, put on hold, or progress the project. If the project is to be progressed, a plan is set for the next stage. This plan must detail the major tasks to be completed and target timescale, and allocate responsibility for completion to appropriate product development personnel.

In all five stages of a product development project, inputs are required from a number of relevant departments within the company: corporate management, finance, sales and marketing, design, development, QC, regulatory affairs, and manufacturing. The designated project manager ensures that appropriate consultation is maintained within the company at all times, and establishes interfaces with customers or external advisors as required.

The concept formation stage aims to assess the validity of an NPD concept. The following issues are considered.

- Target market.
- Relevant standards/regulations.
- Relevant patents.
- Competitive products.
- Manufacturing process.

The initial exploration stage investigates a product or process concept to assess its commercial potential. The areas considered include the following.

- Product and process definition.
- Functional requirement of the product.
- Market price/volume.
- Potential cost/cost savings.
- Competition.
- Existing patents and possible patent protection.
- Manufacturing feasibility.
- Development route and time scale.
- Existing standards and applicable regulations.

If the project is technically and economically feasible, a development program should be planned and completed and, where appropriate, specific budgets are designated and project teams are assigned. The development project should have clearly defined objectives and be planned in some detail. The design inputs and outputs at this stage should be thoroughly documented and regularly reviewed.

The development stage should address the following issues.

- Estimated sales price and volume.
- Possible customers.
- Product performance requirements.
- Prototype product performance.
- Product stability.
- Raw material supplies and specifications.
- In-house manufacturing process and capital cost.
- Subcontract manufacturing options.

- Development agreements.
- Packaging options.
- Regulatory requirements.
- Product safety/toxicity.
- Possible environmental issues.
- Estimated product cost.
- Competitive product performance.
- Intellectual property protection options.
- Product risk analysis.

During the product realization stage, the manufacturing process is scaled up from the pilot to a commercial or semicommercial scale. This can typically happen in four possible ways.

1. Production equipment is set up in house.
2. Technology is licensed to other companies.
3. A joint venture is established.
4. A subcontract manufacturer is sourced.

The output of this stage is typically the following.

- A commissioned manufacturing process.
- A commissioned packaging process.
- A product management system.
- A standard costing.
- Appropriate quality control (QC) and good manufacturing practice procedures.
- Final packaging design.
- Product verified by in vitro, in vivo, or clinical trials as appropriate.

The product launch stage will start when at least one customer agrees to buy the product. The following activities will normally occur during this stage.

- Aging studies of the product in final packaging.
- Packaging integrity validation.
- Raw material source and QC specification agreed with customer.
- In-process QC tests identified.
- Final product specification agreed with customer.
- Product master specification written.
- Packaging specifications written.
- Product risk analysis completed.
- Regulatory submission.
- Product standard cost finalized.
- QC and quality assurance procedures finalized.
- Launch stock produced.
- Suppliers approved and supply agreements signed.
- Distribution/license agreements signed with customer.
- Product transfer price agreed.

The product launch phase is complete when the launch stock has been delivered and accepted by the customer.

16.3 Product development capabilities

A large variety of skills and resources are required to develop a product from concept to regulatory approval and commercial sale. For NPD it has been found that the presence of both functional and integrative capabilities is often positively correlated with NPD process efficiency, measured in terms of lead time, productivity, and product effectiveness. Sustained above-normal returns for a firm can come both from an NPD process that is faster and more productive than competitors', and from new products that create customer value through their overall quality and ability to fit with market needs. As shown in Figure 16.3, in order to develop a product successfully, six types of product development capabilities are required.

To develop a product successfully, six types of product development capabilities are required. First, core technological capabilities are needed to generate the core competitiveness by building quality into the product. Second, technological complementaries, such as previously accumulated technological knowledge, product architecture, and aesthetics, are another source of competitiveness. Third, marketing capabilities are needed to screen, use, and disseminate market information so that market needs can be built into the product. Fourth, marketing policies on pricing, positioning, promotion, etc., are essential ingredients of the design specification. Ancillary services such as testing and clinical trials are also needed to complete the NPD process. Finally, external networks are essential for sourcing external resources and capabilities to complement a firm's internal capabilities.

To build up the capabilities that are needed for developing quality products, medical textile companies need to complement their own capabilities with external sources of capabilities when in-house development lacks economy of scale. More importantly, in a dynamic product market external sources of capabilities are essential, since a fixed set of internal capabilities may not be flexible enough to cope with the pace of market and technology changes. Strategic alliances (SAs) in various forms are therefore essential for the NPD process.

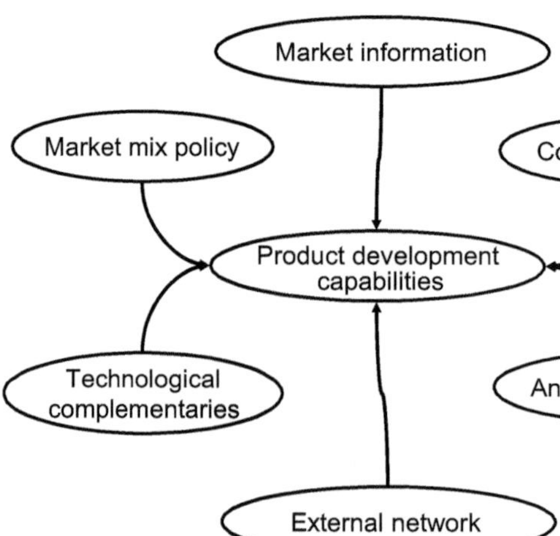

Figure 16.3 The six types of product development capabilities.

SAs can bring in a number of benefits, such as transfer of technology and technological complementary, technology exchange, sharing of research and development (R&D) costs and risks, and access to facilities. There are two basic motives for SAs. The first is cost sharing to share fixed cost, realize economies of scale, and avoid wasteful duplications. The second is skill sharing to benefit from complementary knowledge. For example, firms may have specialized personnel whose skills and knowledge may not be available in the market due to contractual and confidentiality arrangements. Some important NPD results may be withheld from public access. In these cases, SAs provide an important method for outside companies to access a firm's unique resources, which may become the seed of innovation when fused with technologies from the outside firms.

During the development of medical textile products, a firm's technology requirement can be divided into three categories.

1. Core technology: This gives the potential product its unique competitive advantage. For example, in the case of a highly absorbent wound dressing, the core technology is the polymer chemistry that gives the dressing its inherent high-absorption capacity.
2. Complementary technologies: These give the product additional complementary qualities. In the above example, complementary technologies include the manufacturing techniques and packaging design.
3. Ancillary services: These typically fill gaps in the required set of technologies that are needed for developing products, and are often associated with standard testing and trials. In the case of wound dressings, the development work requires a number of ancillary services such as physical and chemical characterizations, biocompatibility testing, and clinical evaluation of the product.

Clearly, the importance of these technologies to a firm's competitive advantage differs in each case. The core technology is often developed in house, since this is a medical textile company's main competitive advantage. To develop quality products with superior advantage over existing products, medical textile companies need to have distinctive core technologies that can give their products an inherent competitiveness. Core technologies are the foundation of medical textile companies and should be developed internally. Indeed, medical textile companies should concentrate their product development efforts around their core technologies.

In addition to the core technologies, the development of medical textile products requires a number of other NPD capabilities which the medical textile companies may not possess. SAs offer an economical way to access these capabilities from external sources. In the search for NPD capabilities that can allow medical textile companies to develop products effectively and efficiently, strategic networking with a diverse range of partners can offer an ideal way of combining the companies' internal capabilities in core technologies with external capabilities in the other important areas of the NPD process.

16.4 Product development strategies

As mentioned previously, product development is in essence a process of providing technical solutions to market needs. In today's world market needs change with the broad environmental changes, while technological solutions can also evolve due to the introduction of new technologies and innovations. A successful product development process needs to match market needs with technical solutions quicker, better, and in a uniquely different way to competitors. To be commercially successful, an effective product development process should be able to embrace the following.

1. Effectively sense market changes.
2. Quickly translate market needs into technical specifications.
3. Deliver superior technical solutions.
4. Have the organizational structure to move fast and efficiently.

Quality of the product and time to market are the two most important objectives of the product development process. This is because buyers and regulatory authorities both put their first priority on product quality, that is, inherent superiority over existing products. Second, the regulatory process is lengthy and technological advances have made the development of nonpatent-infringing "me-too" products much easier. A reduced time to market is therefore important to extend the life span of the medical textile product under development.

A good product development strategy is important for a medical textile company to deliver a high-quality product in a short period of time. This strategy should take into account the company's core competences and make use of its technical and marketing strengths. A key consideration would be whether the company can achieve and maintain a competitive advantage so that the quality of its products and services can be sustained. To achieve these objectives, strategic efforts should be aimed at markets where the company's competitive advantages are the greatest. In particular, strategic considerations should address the following issues.

1. Is the industry attractive?
2. Can the company maintain a large scale and make a long-term commitment?
3. Can competitive advantage be achieved?
4. Are the current company structure, administrative system, and corporate culture suitable to the new competitive environment?

In addition to these considerations, the formulation of a product development strategy should take into consideration the dynamic nature of both the technology and the market for medical textile products. Both market needs and technological solutions change along with other changes in the political, economical, social, and technological aspects of a company's business environment. As can be seen in Figure 16.4, based on the rate of market and technology changes, medical textile products can fall into three types of markets where different product development strategies can be used.

In stable product markets, such as traditional gauzes, strategic management concepts are likely to reflect the emphasis of industrial organizational economics on irreversible commitments to specific assets, hierarchical integration to achieve control of production processes, and defense of established market position. Because neither market preferences for products nor technologies for creating products change significantly, product strategies focus on increasing market share by reducing costs for

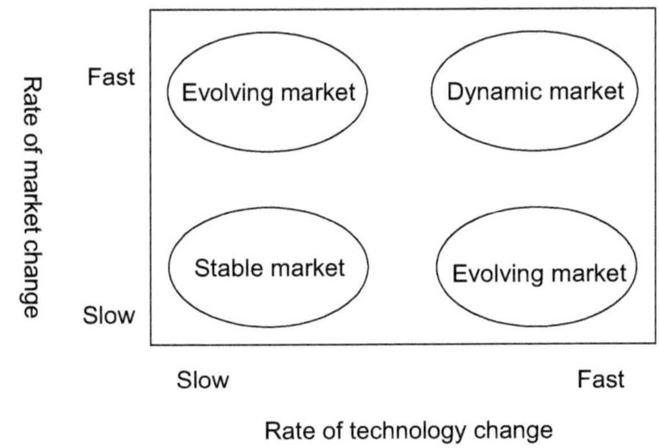

Figure 16.4 Types of markets according to market and technology changes.

producing standard products and extending control of distribution channels. Product differentiation, if attempted at all, is likely to be limited largely to nonproduct dimensions of the offering, such as services, warranties, and advertising. NPD is at most episodic, and when it occurs it is likely to follow a sequential path through the functional divisions that serve the core production and distribution processes of the firm. Vendors of components and production equipment are the principal sources of innovation in products and processes.

In an evolving product market, either technologies or market preferences begin to change. Although technological change can be reflected in either new products or improved production processes, significant market change can also be driven by the evolving preferences of customers, who become increasingly sophisticated in their knowledge about the products. As evolving technologies or market preferences induce changes in market conditions, strategic management concepts focus on strategic adaptation to change, accumulating new resources useful in meeting the new competitive conditions, and reengineering business processes to adjust to new technological and market conditions. The objective of this strategic change process is to acquire the new set of capabilities needed to achieve sustainable competitive advantage in the new competitive context.

Although cost advantages remain important, perceived sources of sustainable competitive advantage are likely to include capabilities that help companies differentiate products by features and performance. Product strategies become driven by forecasting changes in technologies and market preferences. Considerable emphasis will be placed on timing the adoption of new technologies and the introduction of new products that will make obsolete or cannibalize current products. Product development will try to create the new product attributes and the mix of new products thought to be optimal for the new market conditions. Distribution channels may be redesigned to match changes in product offerings better. Product creation processes become driven by market research into evolving customer preferences for new products. To acquire new capabilities in product or process technologies or new market knowledge, the firm may enter into long-term collaborations with other firms that have complementary capabilities in product development, manufacturing, distribution, or marketing. Increased pressure for faster development of new products that incorporate new technologies or serve new customer preferences leads to adoption of multifunctional teams using overlapping problem-solving methods to shorten and improve the development process.

In a dynamic product market, high rates of technological change enable accelerated evolution of product concepts, manufacturing process capabilities, and technologies for coordinating the product creation process. In addition, as customers become more sophisticated in their abilities to identify and select products that better serve their particular needs, customer preferences become much more varied and demanding. The rising number and sophistication of customers create a positive response to firms that offer greater product variety and more frequent introductions of better-performing product models.

As firms search for ways to exploit rapid changes in technologies and market preferences, strategic management concepts should focus on the notions of strategic flexibility, that is, the ability to respond advantageously and quickly to continuous changes in competitive conditions. Strategic management should shift to a process of redesigning the firm as a system for rapidly reconfiguring and redeploying a changing array of assets and capabilities, some of which will be within the firm, but many of which will reside in other firms.

In the face of significant uncertainty about where technologies and market preferences are headed in the long term, accumulating strategic resources that will be sources of sustainable competitive advantage becomes problematic, because the specific resources that will have strategic value in the future cannot be determined with precision. Instead of acquiring specific and increasingly inflexible assets, strategic investments are increasingly channeled into building flexible intellectual assets like human capabilities and knowledge that can be leveraged in a variety of ways as market conditions change. Since no firm can identify or internalize all the intellectual and other assets that may be needed to compete effectively in the future, firms should develop ways of connecting with large numbers of other firms whose various capabilities may prove useful in the future. In essence, firms begin to network with others to improve their abilities to assemble a changing array of resource chains for creating and delivering future new products, which cannot presently be defined with precision.

In terms of the general directions for NPD, medical textile companies should aim to compete on the prices, performance, and functions of their products. As illustrated in Figure 16.5, a cheaper price would be a natural option for those products competing in the stable product market, where customers have consolidated their perceptions about product performance and functions. In an evolving market, technological progress makes it possible to maintain sustained improvement for product qualities and technical performance. Customers are increasingly attracted to products that can offer better and enhanced performance characteristics. Finally, in a dynamic product market, new product features are developed because of the adaptation of new technologies. Products with innovative new functions will be able to differentiate themselves from competitors.

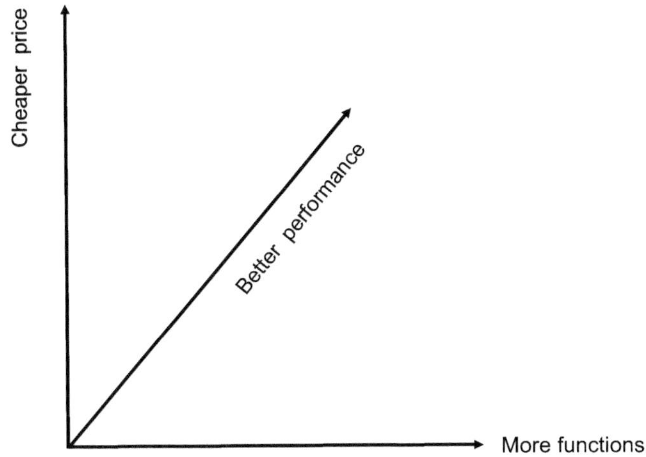

Figure 16.5 The three general directions for new product development.

16.5 Management of a product development project

During the development of a medical textile product, the integration of various functional departments is essential in order to shorten the development time. This includes the interlinking of marketing, R&D, and production activities, and the early integration of customers and suppliers. In the product development arena it is quite common for a design change due to a change in a design input to result in lengthy rework, or even abandonment of the original development plan. A clear definition of the product at an early stage of the NPD process can reduce the need for design change, thereby saving valuable development times.

Medical textile product development is often a complex process involving many subprocesses. For example, to develop a fibrous wound dressing from alginate powders, the powders first need to be made into fibers. The fibers will then be processed into nonwoven felt and cut into individual dressings. Packaging and sterilization processes need to be developed for producing sterile/packed dressings. In addition, test methods need to be developed to control the processes and provide data for regulatory approval. In such a situation, managing development time is a combination of managing a specific functional process, such as the fiber making, and overall project management involving all the subprocesses. Cutting development time involves the streamlining of each process while undertaking all the processes in parallel.

In addition to managing development time, a major issue in the management of the NPD process is to convert organizational resources in their various forms, such as financial, human, and technological, into NPD capabilities. In the NPD process, both process efficiency in terms of the time and cost of development and product effectiveness in terms of the fit to market need and product efficacy are affected by the behavior of different agents, including team members, project leaders, senior managers, customers, and suppliers. For example, employing a project leader with internal management skills can by itself lead to an increase in product concept effectiveness, thanks to his or her ability to include team members' creativity.

In a rapidly changing environment, the ability of people to work together has become the cornerstone ingredient of organizational success. In an NPD environment, no individual R&D professional can possibly understand the entire design, development, production, marketing, distribution, and service processes. Nor can any single R&D professional have all the requisite knowledge needed to do his or her job. To be both timely and effective, R&D individuals have to be able to integrate their insights, creativity, and accomplishments with each other and with the demands and ongoing activities of other relevant organizational constituents.

However, although the need for teamwork is compelling, practical experience often shows that managing R&D professionals is very much like herding cats—that is, they tend to behave rather independently and are not especially responsive to or tolerant of their organization's formal rules, procedures, and bureaucratic demands. Teamwork in these situations can be challenging, and involves skilled management of both technologies and people.

The R&D and marketing interface is of critical importance in the NPD process. Traditionally, this interface often has problems such as a lack of communication and appreciation, mutual distrust, and inadequate constructive criticism. To establish a good interface, attempts should be made to involve both R&D and marketing at early stages of the NPD process and create formal organizational structures, such as developing protocols and agreement on each side's roles and responsibilities in the product development process. In a dynamic product market where technologies and market preferences change rapidly, the flows of knowledge and information become particularly important. Cross-functional reporting procedures, the representation of multiple disciplines in decision making, and the appointment of a heavyweight product manager with authority over functional specialists are effective ways to reduce information asymmetries.

At the technical level, medical textile product development often involves a multidisciplinary approach, beginning with a theoretical concept of the product. As shown in Figure 16.6, the technical development of a medical textile product proceeds from

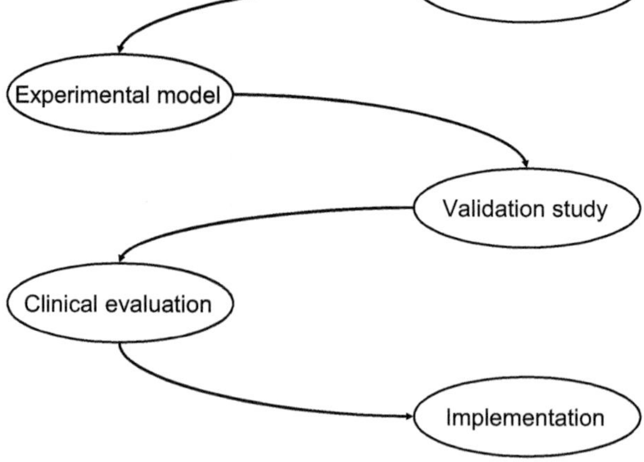

Figure 16.6 The process for the research and development of medical textile materials.

the formation of a theoretical concept to the full implementation of the product, going through studies on the experimental model, validation and verification of prototype samples, and clinical evaluation of products prior to regulatory approval from appropriate competent authorities. Because of the intensive demand for knowledge and skills from many different disciplines, the development of medical textile products often requires a large collaboration between industrial partners ranging from multinational corporations to small- and medium-sized enterprises, clinicians, and academics.

16.6 Summary

Medical textile product development is a unique business process that requires unique management practices at the strategic, business, and project levels. At the strategic level, the social, economical, political, and technological changes in the world have presented medical textile companies with many business opportunities. This is because while the aging population provides an increasing demand for new and effective products, the emergence of new technologies has generated more and more technical possibilities.

The business environment also presents medical textile companies with a number of potential threats. As governments face escalating healthcare costs, they are increasingly demanding high-quality products at low cost. At the same time, development costs are rising because of quality requirements and the lengthy process of regulatory approval. The short product lifecycle, the high cost of development, and the need to launch products globally all demand a speedy development process for medical textile products. To manage this, medical textile companies need to concentrate their NPD activities around their core technologies and key markets. In addition, they should build strategic networks to access the external sources of capabilities that are needed to complement their internal capabilities in NPD.

Overall, the successful development of medical textile products depends on a number of key success factors, which include the following.

1. Possession of proprietary core technologies.
2. Access to knowledge networks.
3. Alliances with major healthcare firms and research organizations.
4. A product development process which can shorten the time from concept formation to regulatory approval.
5. An organizational structure which is responsive to internal and external changes.
6. A management system with emphasis on teamwork and communications.

Sources for further information

[1] Loch C, Kavadias S. Handbook of new product development management. New York: Butterworth Heinemann; 2007.
[2] Cooper RG, Kleinschmidt EJ. New product success factors: a comparison of kills versus successes and failures. R&D Manage 1990;20(1):47–52.

[3] Dibb S, Russell Y, Simkin L. The EU marketing environment: pharmaceuticals and Japanese strategy. Eur Manage J 1997;15:195–9.
[4] Anderson J. Technology foresight for competitive advantage. Long Range Plann 1997;30:665–9.
[5] Sapienza AM. R&D collaboration as a global competitive tactic-biotechnology and the ethical pharmaceutical industry. R&D Manage 1989;19(4):285–9.
[6] Cooper AC. Research findings in strategic management with implications for R&D management. R&D Manage 1989;19(2):115–20.
[7] Verona G. A resource based view of product development. Acad Manage Review 1999;24(1):132–6.
[8] Forrest JE, Martin JC. Strategic alliances between large and small research intensive organisations: experiences in the biotechnology industry. R&D Manage 1992;22(1):41–7.
[9] Scott GM. The new age of new product development: are we there yet? R&D Manage 1998;28:225–31.
[10] Berry M. Strategic planning in small high tech companies. Long Range Plann 1998;31:455–61.
[11] Sanchez R. Strategic product creation: managing new interactions of technology, markets and organisations. Eur Manage J 1996;14:121–9.
[12] Topfer A. New product-cutting the time to market. Long Range Plann 1995;28:61–6.
[13] Towner SJ. Four ways to accelerate new product development. Long Range Plann 1994;27:57–61.
[14] Cohen MA, Eliashberg J, Ho T. NPD: the performance and time to market trade-off. Manage Sci 1996;42:173–7.
[15] Gupta AK, Wilemon D. Improving R&D marketing relations: R&D's perspective. R&D Manage 1990;20(4):277–82.
[16] Cooper R. Benchmarking new product performance: results of the best practice study. Eur Manage J 1998;16:1–7.
[17] Nellore R, Soderquist K, Eriksson K. A specification model for product development. Eur Manage J 1999;17:50–5.
[18] Browne N, Grocott P, Cowley S, et al. Woundcare research for appropriate products (WRAP): validation of the TELER method involving users. Int J Nurs Stud 2004;41(5):559–71.

Index

Note: Page numbers followed by "f" and "t" indicate figures and tables, respectively.

M

P